And others, William Francis Magie

The Second Law of Thermodynamics

memoirs by Carnot, Clausius and Thompson

And others, William Francis Magie

The Second Law of Thermodynamics
memoirs by Carnot, Clausius and Thompson

ISBN/EAN: 9783337232627

Printed in Europe, USA, Canada, Australia, Japan

Cover: Foto ©berggeist007 / pixelio.de

More available books at **www.hansebooks.com**

THE SECOND LAW OF

THERMODYNAMICS

MEMOIRS BY CARNOT, CLAUSIUS
AND THOMSON

TRANSLATED AND EDITED

By W. F. MAGIE, Ph.D.

PROFESSOR OF PHYSICS IN PRINCETON UNIVERSITY

NEW YORK AND LONDON
HARPER & BROTHERS PUBLISHERS
1899

PREFACE

AFTER the invention of the steam-engine in its present form by James Watt, the attention of engineers and of scientific men was directed to the problem of its further improvement. With this end in view, the young Sadi Carnot, in 1824, published the *Réflexions sur la Puissance Motrice du Feu*, of which the translation is given in this volume. In this really great memoir, Carnot examined the relations between heat and the work done by heat used in an ideal engine, and by reducing the problem to its simplest form and avoiding all special questions relating to details, he succeeded in establishing the conditions upon which the economical working of all heat-engines depends. It is not necessary here to animadvert upon the use made by Carnot of the substantial theory of heat, and the consequent failure of the proof of his main proposition when the true nature of heat was appreciated. It is sufficient to say that though the proof was invalid, the proposition remained true, and carried with it the truth of such of Carnot's deductions as were based solely upon it.

Carnot's memoir remained for a long time unappreciated, and it was not until use was made of it by William Thomson (now Lord Kelvin), in 1848, to establish an absolute scale of temperature, that the merits of the method proposed in it were recognized. In his first paper on this subject Thomson retained the substantial theory of heat, but the evidence in favor of the mechanical theory became so strong that he soon after adopted the new view. Applying it to the questions treated by Carnot, he found that Carnot's proposition could no longer be proved by denying the possibility of " the perpetual motion," and was led to lay down a second fundamental principle to serve in the demonstration. This principle is now called the Second Law of Thermodynamics. A part of the memoir in which this principle is stated and many of its consequences developed is given in this volume. It was published in March, 1851.

In the previous year Clausius published a discussion of the same question as that treated by Thomson, in which he lays down a principle for use in the demonstration of Carnot's proposition, which, while not the same in form as Thomson's, is the same in content, and ranks as another statement of the Second Law of Thermodynamics. His paper is also given in this volume. While not so powerful or so inclusive as Thomson's, it deserves attention for the clearness and simplicity of its form. Clausius followed up this paper by others, and subsequently published a book in which the subject of Thermodynamics was given a systematic treatment, and in which he introduced and developed the important function called by him the entropy.

The science of Thermodynamics, founded by the labors of these three illustrious men, has led to the most important developments in all departments of physical science. It has pointed out relations among the properties of bodies which could scarcely have been anticipated in any other way ; it has laid the foundation for the Science of Chemical Physics ; and, taken in connection with the kinetic theory of gases, as developed by Maxwell and Boltzmann, it has furnished a general view of the operations of the universe which is far in advance of any that could have been reached by purely dynamical reasoning.

GENERAL CONTENTS

REFLECTIONS ON
THE MOTIVE POWER OF HEAT

BY

SADI CARNOT

Paris, 1824

CONTENTS

REFLECTIONS ON
THE MOTIVE POWER OF HEAT AND
ON ENGINES SUITABLE FOR
DEVELOPING THIS POWER

BY

SADI CARNOT

IT is well known that heat may be used as a cause of motion, and that the motive power which may be obtained from it is very great. The steam-engine, now in such general use, is a manifest proof of this fact.

To the agency of heat may be ascribed those vast disturbances which we see occurring everywhere on the earth ; the movements of the atmosphere, the rising of mists, the fall of rain and other meteors,* the streams of water which channel the surface of the earth, of which man has succeeded in utilizing only a small part. To heat are due also volcanic eruptions and earthquakes. From this great source we draw the moving force necessary for our use. Nature, by supplying combustible material everywhere, has afforded us the means of generating heat and the motive power which is given by it, at all times and in all places, and the steam-engine has made it possible to develop and use this power.

The study of the steam-engine is of the highest interest, owing to its importance, its constantly increasing use, and the great changes it is destined to make in the civilized world. It has already developed mines, propelled ships, and dredged rivers and harbors. It forges iron, saws wood, grinds grain, spins and weaves stuffs, and transports the heaviest loads. In the future it will most probably be the universal motor, and

* [*Any atmospheric phenomenon was formerly called a meteor.*]

3

will furnish the power now obtained from animals, from water-falls, and from air-currents. Over the first of these motors it has the advantage of economy, and over the other two the incalculable advantage that it can be used everywhere and always, and that its work need never be interrupted. If in the future the steam-engine is so perfected as to render it less costly to construct it and to supply it with fuel, it will unite all desirable qualities and will promote the development of the industrial arts to an extent which it is difficult to foresee. It is, indeed, not only a powerful and convenient motor, which can be set up or transported anywhere, and substituted for other motors already in use, but it leads to the rapid extension of those arts in which it is used, and it can even create arts hitherto unknown.

The most signal service which has been rendered to England by the steam-engine is that of having revived the working of her coal-mines, which had languished and was threatened with extinction on account of the increasing difficulty of excavation and extraction of the coal.* We may place in the second rank the services rendered in the manufacture of iron, as much by furnishing an abundant supply of coal, which took the place of wood as the wood began to be exhausted, as by the powerful machines of all kinds the use of which it either facilitated or made possible.

Iron and fire, as every one knows, are the mainstays of the mechanical arts. Perhaps there is not in all England a single industry whose existence is not dependent on these agents, and which does not use them extensively. If England were to-day to lose its steam-engines it would lose also its coal and iron, and this loss would dry up all its sources of wealth and destroy its prosperity; it would annihilate this colossal power. The destruction of its navy, which it considers its strongest support, would be, perhaps, less fatal.

The safe and rapid navigation by means of steamships is an

---

* One may safely say that the mining of coal has increased tenfold since the invention of the steam-engine. The mining of copper, of tin, and of iron has increased almost as much. The effect produced half a century ago in the mines of England is now being repeated in the gold and silver mines of the New World, the working of which was steadily declining, principally on account of the insufficiency of the motors used for the excavation and extraction of the minerals.

entirely new art due to the steam-engine. This art has already made possible the establishment of prompt and regular communication on the arms of the sea, and on the great rivers of the old and new continents. By means of the steam-engine regions still savage have been traversed which but a short time ago could hardly have been penetrated. The products of civilization have been taken to all parts of the earth, which they would otherwise not have reached for many years. The navigation due to the steam-engine has in a measure drawn together the most distant nations. It tends to unite the peoples of the earth as if they all lived in the same country. In fact, to diminish the duration, the fatigue, the uncertainty and danger of voyages is to lessen their length.*

The discovery of the steam-engine, like most human inventions, owes its birth to crude attempts which have been attributed to various persons and of which the real author is not known. The principal discovery consists indeed less in these first trials than in the successive improvements which have brought it to its present perfection. There is almost as great a difference between the first structures where expansive force was developed and the actual steam-engine as there is between the first raft ever constructed and a man-of-war.

If the honor of a discovery belongs to the nation where it acquired all its development and improvement, this honor cannot in this case be withheld from England: Savery, Newcomen, Smeaton, the celebrated Watt, Woolf, Trevithick, and other English engineers, are the real inventors of the steam-engine. At their hands it received each successive improvement. It is natural that an invention should be made, improved, and perfected where the need of it is most strongly felt.

In spite of labor of all sorts expended on the steam-engine, and in spite of the perfection to which it has been brought, its theory is very little advanced, and the attempts to better this state of affairs have thus far been directed almost at random.

The question has often been raised whether the motive power

* We speak of diminishing the danger of voyages ; in fact, though the use of the steam-engine in ships is attended with some dangers, these are always exaggerated and are compensated for by the ability of ships to keep a definite course, and to resist winds which would otherwise drive the vessel on the coast, or on shoals or reefs.

of heat is limited or not;* whether there is a limit to the possible improvements of the steam-engine which, in the nature of the case, cannot be passed by any means; or if, on the other hand, these improvements are capable of indefinite extension. Inventors have tried for a long time, and are still trying, to find whether there is not a more efficient agent than water by which to develop the motive power of heat; whether, for example, atmospheric air does not offer great advantages in this respect. We propose to submit these questions to a critical examination.

The phenomenon of the production of motion by heat has not been considered in a sufficiently general way. It has been treated only in connection with machines whose nature and mode of action do not admit of a full investigation of it. In such machines the phenomenon is, in a measure, imperfect and incomplete; it thus becomes difficult to recognize its principles and study its laws. To examine the principle of the production of motion by heat in all its generality, it must be conceived independently of any mechanism or of any particular agent; it is necessary to establish proofs applicable not only to steam-engines† but to all other heat-engines, irrespective of the working substance and the manner in which it acts.

The machines which are not worked by heat—for instance, those worked by men or animals, by water-falls, or by air currents—can be studied to their last details by the principles of mechanics. All possible cases may be anticipated, all imaginable actions are subject to general principles already well established and applicable in all circumstances. The theory of such machines is complete. Such a theory is evidently lacking for heat-engines. We shall never possess it until the laws of physics are so extended and generalized as to make known in advance all the effects of heat acting in a definite way on any body whatsoever.

---

* The expression motive power here signifies the useful effect that a motor is capable of producing. This effect may always be measured in terms of the elevation of a weight through a certain distance; it is measured, as is well known, by the product of the weight and the height to which it is raised.

† We distinguish here between the steam-engine and the heat-engine in general, which can be worked by any agent, and not by water vapor only, to realize the motive power of heat.

We shall take for granted in what follows a knowledge, at least a superficial one, of the various parts which compose an ordinary steam-engine. We think it unnecessary to describe the fire-box, the boiler, the steam-chest, the piston, the condenser, etc.

The production of motion in the steam-engine is always accompanied by a circumstance which we should particularly notice. This circumstance is the re-establishment of equilibrium in the caloric*—that is, its passage from one body where the temperature is more or less elevated to another where it is lower. What happens, in fact, in a steam-engine at work? The caloric developed in the fire-box as an effect of combustion passes through the wall of the boiler and produces steam, incorporating itself with the steam in some way. This steam, carrying the caloric with it, transports it first into the cylinder, where it fulfils some function, and thence into the condenser, where the steam is precipitated by coming in contact with cold water. As a last result the cold water in the condenser receives the caloric developed by combustion. It is warmed by means of the steam, as if it had been placed directly on the fire-box. The steam is here only a means of transporting caloric; it thus fulfils the same office as in the heating of baths by steam, with the exception that in the case in hand its motion is rendered useful.

We can easily perceive, in the operation which we have just described, the re-establishment of equilibrium in the caloric and its passage from a hotter to a colder body. The first of these bodies is the heated air of the fire-box; the second, the water of condensation. The re-establishment of equilibrium of the caloric is accomplished between them—if not completely, at least in part; for, on the one hand, the heated air after having done its work escapes through the smoke-stack at a much lower temperature than that which it had acquired by the combustion; and, on the other hand, the water of the condenser, after having precipitated the steam, leaves the engine with a higher temperature than that which it had when it entered.

The production of motive power in the steam-engine is

---

* [*Caloric is heat considered as an indestructible substance. The word is used by Carnot interchangeably with feu, fire, or heat.*]

7

therefore not due to a real consumption of the caloric, *but to its transfer from a hotter to a colder body*—that is to say, to the re-establishment of its equilibrium, which is assumed to have been destroyed by a chemical action such as combustion, or by some other cause. We shall soon see that this principle is applicable to all engines operated by heat.

According to this principle, to obtain motive power it is not enough to produce heat ; it is also necessary to provide cold, without which the heat would be useless. For if there existed only bodies as warm as our furnaces, how would the condensation of steam be possible, and where could it be sent if it were once produced? It cannot be replied that it could be ejected into the atmosphere, as is done with certain engines,* since the atmosphere would not receive it. In the actual state of things the atmosphere acts as a vast condenser for the steam, because it is at a lower temperature ; otherwise it would soon be saturated, or, rather, would be saturated in advance.†

Everywhere where there is a difference of temperature, and where the re-establishment of equilibrium of the caloric can be effected, the production of motive power is possible. Water vapor is one agent for obtaining this power, but it is not the only one ; all natural bodies can be applied to this purpose, for they are all susceptible to changes of volume, to successive contractions and dilatations effected by alternations of heat and cold ; they are all capable, by this change of volume, of overcoming resistances and thus of developing motive power. A

---

* Some high-pressure engines eject vapor into the atmosphere instead of condensing it. They are used mostly in places where it is difficult to procure a current of cold water sufficient to effect condensation.

† The existence of water in a liquid state, which is here necessarily assumed, since without it the steam-engine could not be supplied, presupposes the existence of a pressure capable of preventing it from evaporating, and consequently of a pressure equal to or greater than the tension of the vapor at the temperature of the water. If such a pressure were not exerted by the atmosphere a quantity of water vapor would instantly be produced sufficient to exert this pressure on itself, and this pressure must always be overcome in ejecting the steam of the engine into the new atmosphere. This is evidently equivalent to overcoming the tension which is exerted by the vapor after it has been condensed by the ordinary means.

If a very high temperature were to prevail at the surface of the earth, as it almost certainly does in its interior, all the water of the oceans would exist in the form of vapor in the atmosphere, and there would be no water in a liquid state.

solid body, such as a metallic bar, when alternately heated and cooled, increases and diminishes in length and can move bodies fixed at its extremities. A liquid, alternately heated and cooled, increases and diminishes in volume and can overcome obstacles more or less great opposed to its expansion. An aeriform fluid undergoes considerable changes of volume with changes of temperature ; if it is enclosed in an envelope capable of enlargement, such as a cylinder furnished with a piston, it will produce movements of great extent. The vapors of all bodies which are capable of evaporation, such as alcohol, mercury, sulphur, etc., can perform the same function as water vapor. This, when alternately heated and cooled, will produce motive power in the same way as permanent gases, without returning to the liquid state. Most of these means have been proposed, several have been even tried, though, thus far, without much success.

We have explained that the motive power in the steam-engine is due to a re-establishment of equilibrium in the caloric ; this statement holds not only for steam-engines but also for all heat-engines—that is to say, for all engines in which caloric is the motor. Heat evidently can be a cause of motion only through the changes of volume or of form to which it subjects the body ; those changes cannot occur at a constant temperature, but are due to alternations of heat and cold ; thus to heat any substance it is necessary to have a body warmer than it, and to cool it, one cooler than it. We must take caloric from the first of these bodies and transfer it to the second by means of the intermediate body, which transfer re-establishes, or, at least, tends to re-establish, equilibrium of the caloric.

At this point we naturally raise an interesting and important question : Is the motive power of heat invariable in quantity, or does it vary with the agent which one uses to obtain it— that is, with the intermediate body chosen as the subject of the action of heat ?

It is clear that the question thus raised supposes given a certain quantity of caloric * and a certain difference of temperature.

---

* It is unnecessary to explain here what is meant by a quantity of caloric or of heat (for we use the two expressions interchangeably), or to describe how these quantities are measured by the calorimeter ; nor shall we explain the terms latent heat, degree of temperature, specific heat, etc. The reader should be familiar with these expressions from his study of the elementary treatises of physics or chemistry.

For example, we suppose that we have at our disposal a body, *A*, maintained at the temperature 100 degrees, and another body, *B*, at 0 degrees, and inquire what quantity of motive power will be produced by the transfer of a given quantity of caloric— for example, of so much as is necessary to melt a kilogram of ice—from the first of these bodies to the second; we inquire if this quantity of motive power is necessarily limited; if it varies with the substance used to obtain it; if water vapor offers in this respect more or less advantage than vapor of alcohol or of mercury, than a permanent gas or than any other substance. We shall try to answer these questions in the light of the considerations already advanced.

We have previously called attention to the fact, which is self-evident, or at least becomes so if we take into consideration the changes of volume occasioned by heat, that *wherever there is a difference of temperature the production of motive power is possible.* Conversely, wherever this power can be employed, it is possible to produce a difference of temperature or to destroy the equilibrium of the caloric. Percussion and friction of bodies are means of raising their temperature spontaneously* to a higher degree than that of surrounding bodies, and consequently of destroying that equilibrium in the caloric which had previously existed. It is an experimental fact that the temperature of gaseous fluids is raised by compression and lowered by expansion. This is a sure method of changing the temperature of bodies, and thus of destroying the equilibrium of the caloric in the same substance, as often as we please. Steam, when used in a reverse way from that in which it is used in the steam-engine, can thus be considered as a means of destroying the equilibrium of the caloric. To be convinced of this, it is only necessary to notice attentively the way in which motive power is developed by the action of heat on water vapor. Let us consider two bodies, *A* and *B*, each maintained at a constant temperature, that of *A* being higher than that of *B*; these two bodies, which can either give up or receive heat without a change of temperature, perform the functions of two indefinitely great reservoirs of caloric. We will call the first body the source and the second the refrigerator.

If we desire to produce motive power by the transfer of a

* [*That is, without the communication of heat.*]

certain quantity of heat from the body A to the body B we may proceed in the following way:

1. We take from the body A a quantity of caloric to make steam—that is, we cause A to serve as the fire-pot, or rather as the metal of the boiler in an ordinary engine; we assume the steam produced to be at the same temperature as the body A.

2. The steam is received into an envelope capable of enlargement, such as a cylinder furnished with a piston. We then increase the volume of this envelope, and consequently also the volume of the steam. The temperature of the steam falls when it is thus rarefied, as is the case with all elastic fluids; let us assume that the rarefaction is carried to the point where the temperature becomes precisely that of the body B.

3. We condense the steam by bringing it in contact with B and exerting on it at the same time a constant pressure until it becomes entirely condensed. The body B here performs the function of the injected water in an ordinary engine, with the difference that it condenses the steam without mixing with it and without changing its own temperature.* The operations which we have just described could have been performed in a reverse sense and order. There is nothing to prevent the for-

---

* It will perhaps excite surprise that B, being at the same temperature as the steam, can condense it. Without doubt this is not rigorously possible, but the smallest difference in temperature will determine condensation. This remark is sufficient to establish the propriety of our reasoning. In the same way, in the differential calculus, to obtain an exact result it is sufficient to be able to conceive of the quantities neglected as capable of being indefinitely diminished relative to the quantities retained in the equation.

The body B condenses the steam without changing its own temperature. We have assumed that this body is maintained at a constant temperature. The caloric is therefore taken from it as fast as it is given up to it by the steam. An example of such a body is furnished by the metallic walls of the condenser when the vapor is condensed in it by means of cold water applied to the outside, as is done in some engines. In the same way the water of a reservoir can be maintained at a constant level, if the liquid runs out at one side as fast as it comes in at the other.

One could even conceive the bodies A and B such that they would remain of themselves at a constant temperature though losing or gaining quantities of heat. If, for example, the body A were a mass of vapor ready to condense and the body B a mass of ice ready to melt, these bodies, as is well known, could give out or receive caloric without changing their temperature.

11

mation of vapor by means of the caloric of the body B, and its compression from the temperature of B, in such a way that it acquires the temperature of the body A, and then its condensation in contact with A, under a pressure which is maintained constant until it is completely liquefied.

In the first series of operations there is at the same time a production of motive power and a transfer of caloric from the body A to the body B; in the reverse series there is at the same time an expenditure of motive power and a return of the caloric from B to A. But if in each case the same quantity of vapor has been used, if there is no loss of motive power or of caloric, the quantity of motive power produced in the first case will equal the quantity expended in the second, and the quantity of caloric which in the first case passed from A to B will equal the quantity which in the second case returns from B to A, so that an indefinite number of such alternating operations can be effected without the production of motive power or the transfer of caloric from one body to the other. Now if there were any method of using heat preferable to that which we have employed, that is to say, if it were possible that the caloric should produce, by any process whatever, a larger quantity of motive power than that produced in our first series of operations, it would be possible, by diverting a portion of this power, to effect a return of caloric, by the method just indicated, from the body B to the body A—that is, from the refrigerator to the source—and thus to re-establish things in their original state, and to put them in position to recommence an operation exactly similar to the first one, and so on : there would thus result not only the perpetual motion, but an indefinite creation of motive power without consumption of caloric or of any other agent whatsoever. Such a creation is entirely contrary to the ideas now accepted, to the laws of mechanics and of sound physics; it is inadmissible.* We may hence con-

---

* The objection will perhaps here be made that perpetual motion has only been demonstrated to be impossible in the case of mechanical actions, and that it may not be so when we employ the agency of heat or of electricity ; but can we conceive of the phenomena of heat and of electricity as due to any other cause than some motion of bodies, and, as such, should they not be subject to the general laws of mechanics ? Besides, do we not know *a posteriori* that all the attempts made to produce perpetual motion by any means whatever have been fruitless ; that no truly perpetual motion

clude that *the maximum motive power resulting from the use of steam is also the maximum motive power which can be obtained by any other means.* We shall soon give a second and more rigorous demonstration of this law. What has been given should only be regarded as a sketch (see page 15).

It may properly be asked, in connection with the proposition just stated, what is the meaning of the word *maximum?* How can we know that this maximum is reached and that the steam is used in the most advantageous way possible to produce motive power?

Since any re-establishment of equilibrium in the caloric can be used to produce motive power, any re-establishment of equilibrium which is effected without producing motive power should be considered as a veritable loss: now, with little reflection, we can see that any change of temperature which is not due to a change of volume of the body can be only a useless re-establishment of equilibrium in the caloric.* The necessary condition of the maximum is, then, that *in bodies used to obtain the motive power of heat, no change of temperature occurs which is*

has ever been produced, meaning by that, a motion which continues indefinitely without change in the body used as an agent?

The electromotive apparatus (Volta's pile) has sometimes been considered capable of producing perpetual motion; the attempt has been made to realize it by the construction of the dry pile, which is claimed to be unalterable; but, in spite of all that has been done, the apparatus always deteriorates perceptibly when its action is sustained for some time with any energy.

The general and philosophical acceptation of the words *perpetual motion* should comprehend not only a motion capable of indefinite continuance after it has been started, but also the action of an apparatus, of a set of bodies, capable of creating motive power in an unlimited quantity, and of setting in motion successively all the bodies of nature, if they are originally at rest, and of destroying in them the principle of inertia, and finally capable of furnishing in itself all the forces necessary to move the entire universe, to prolong and to constantly accelerate its motion. Such would be a real creation of motive power. If this were possible, it would be useless to search for motive power in combustibles, in currents of water and air. We should have at our disposal an inexhaustible source from which we could draw at will.

* We do not here take into consideration any chemical action between the bodies used to obtain the motive power of heat. The chemical action which occurs in the source is in a sense preliminary, an action not designed to immediately create motive power, but to destroy equilibrium in the caloric, to produce a difference in temperature which shall finally result in motion.

not due to a change of volume. Conversely, every time that this condition is fulfilled, the maximum is attained.

This principle should not be lost sight of in the construction of heat-engines. It is the foundation upon which they rest. If it cannot be rigorously observed, it should at least be departed from as little as possible.

Any change of temperature which is not due to a change of volume or to chemical action (which we provisionally assume not to occur in this case) is necessarily due to the direct transfer of caloric from a hotter to a colder body. This transfer takes place principally at the points of contact of bodies at different temperatures; thus such contacts should be avoided as much as possible. They doubtless cannot be avoided entirely, but at least care should be taken that the bodies brought in contact should differ but little in temperature.

When we assumed in the previous demonstration that the caloric of the body *A* was used to produce steam, we supposed the steam to be produced at the same temperature as that of the body *A*; thus the only contact was between two bodies of equal temperature; the change of temperature which the steam afterwards experienced was due to expansion and consequently to a change of volume; finally condensation was effected without contact of bodies of different temperatures. It was effected by the exercise of a constant pressure on the steam brought in contact with the body *B*, at the same temperature as that of the body *B*. The condition of the maximum was thus fulfilled. In reality things would not occur exactly as we have supposed. In order to effect a transfer of the caloric from one body to the other, the first must have the higher temperature ; but this difference may be supposed to be as small as we please ; we may, in theory, consider it zero without invalidating the argument.

A more valid objection may be made to our demonstration, namely :

When we produce steam by taking caloric from the body *A*, and when this steam is afterwards condensed by contact with the body *B*, the water used to form it, which was assumed to be, at the beginning, at the temperature of the body *A*, is, at the end of the operation, at the temperature of the body *B*—that is, it is colder. If we wish to recommence an operation similar to the first, to develop a new quantity of motive power with the same instrument and the same steam, we must first re-establish the original state of things and bring the water to the temperature which it had at first. This can no doubt be done by placing it immediately in contact with the body A; but in that case there is contact between bodies of different temperatures and loss of motive power.* It would become impossible to perform the reverse operation—that is, to cause the caloric used in raising the temperature of the liquid to return to the body A.

This difficulty can be removed by supposing the difference of temperature between the body A and the body B infinitely small; the quantity of heat needed to bring the liquid back to its original temperature is also infinitely small and negligible relatively to that finite quantity which is needed to produce the steam.

The proposition being thus demonstrated for the case in which the difference of temperature of the two bodies is infinitely small may easily be extended to cover the general case. In fact, if we desire to produce motive power by the transfer of caloric from the body A to the body Z, the temperature of the latter body being very different from that of the former, we may imagine a series of bodies B, C, D . . . at temperatures intermediate between those of the bodies A and Z, and chosen in such a manner that the differences between A and B, B and C . . . shall be always infinitely small. The caloric which proceeds from A arrives at Z only after having passed through the bodies B, C, D . . . and after having developed in each of these transfers the maximum of motive power. The reverse operations are here all possible, and the reasoning on page 11 becomes rigorously applicable.

According to the views now established we may with pro-

---

* This kind of loss is always met with in steam-engines. In fact, the water which supplies the boiler is always colder than that which it already contains, and hence a useless re-establishment of equilibrium in the caloric takes place between them. It is easy to see *a posteriori* that this re-establishment of equilibrium entails a loss of motive power if we reflect that it would be possible to heat the water supply before injecting it by using it as water of condensation in a small accessory engine, in which steam taken from the large boiler could be used and in which condensation would occur at a temperature intermediate between that of the boiler and that of the principal condenser. The force produced by the small engine would entail no expenditure of heat, since all that it would use would re-enter the boiler with the water of condensation.

priety compare the motive power of heat with that of a water-fall; both have a maximum which cannot be surpassed, whatever may be, on the one hand, the machine used to receive the action of the water and whatever, on the other hand, the substance used to receive the action of the heat. The motive power of falling water depends on the quantity of water and on the height of its fall; the motive power of heat depends also on the quantity of caloric employed and on that which might be named, which we, in fact, will call, *its descent**—that is to say, on the difference of temperature of the bodies between which the exchange of caloric is effected. In the fall of water the motive power is strictly proportional to the difference of level between the higher and lower reservoirs. In the fall of caloric the motive power doubtless increases with the difference of temperature between the hotter and colder bodies, but we do not know whether it is proportional to this difference. We do not know, for example, whether the fall of the caloric from 100 to 50 degrees furnishes more or less motive power than the fall of the same caloric from 50 degrees to zero. This is a question which we propose to examine later.

We shall give here a second demonstration of the fundamental proposition stated on page 13 and present this proposition in a more general form than we have before.

When a gaseous fluid is rapidly compressed its temperature rises, and when it is rapidly expanded its temperature falls. This is one of the best established facts of experience; we shall take it as the basis of our demonstration.† When the temperature of a gas is raised and we wish to bring it back to its

---

* The matter here treated being entirely new, we are obliged to employ expressions hitherto unused and which are not perhaps as clear as could be desired.

† The facts of experience which best prove the change of temperature of a gas by compression or expansion are the following :

1. The fall of temperature indicated by a thermometer placed under the receiver of an air-pump in which a vacuum is produced. This is very perceptible with a Bréguet thermometer ; it may amount to upwards of 40 or 50 degrees. The cloud which is formed in this operation seems to be due to the condensation of water vapor caused by the cooling of the air.

2. The ignition of tinder in the so-called fire-syringe (pneumatic tinder-box), which is, as is well known, a small pump in which air may be rapidly compressed.

3. The fall of temperature indicated by a thermometer placed in a re-

original temperature without again changing its volume, it is necessary to remove caloric from it. This caloric may also be removed as the compression is effected, so that the temperature of the gas remains constant. In the same way, if the gas is rarefied, we can prevent its temperature from falling, by furnishing it with a certain quantity of caloric. We shall call the caloric used in such cases, when it occasions no change of temperature, *caloric due to a change of volume.* This name does not indicate that the caloric belongs to the volume; it does not belong to it any more than it does to the pressure, and it might equally well be called *caloric due to a change of pressure.* We are ignorant of what laws it obeys in respect to changes of volume: it is possible that its quantity changes with the nature of the

ceptacle in which air has been compressed, and from which it is allowed to escape by opening a stopcock.

4. The results of experiments on the velocity of sound. M. de Laplace has shown that to harmonize these results with theory and calculation we must assume that air is heated by a sudden compression.

The only fact which can be opposed to these is an experiment of MM. Gay-Lussac and Welter, described in the *Annales de Chimie et de Physique.* If a small opening is made in a large reservoir of compressed air, and the bulb of·a thermometer is placed in the current of air escaping through this opening, no perceptible fall of temperature is indicated by the thermometer.

We may explain this fact in two ways :

1. The friction of the air against the walls of the opening through which it escapes may perhaps develop enough heat to be noticed ; 2. The air which impinges immediately upon the bulb of the thermometer may recover by its shock against the bulb, or rather by the détour which it is forced to make by the encounter, a density equal to that which it had in the receiver, somewhat in the same way as a current of water rises above its level when it meets a fixed obstacle.

The change of temperature in gases occasioned by a change of volume may be considered one of the most important facts in physics, because of the innumerable consequences which it entails, and at the same time as one of the most difficult to elucidate and to measure by conclusive experiments. It presents singular anomalies in several cases.

Must we not attribute the coldness of the air in high regions of the atmosphere to the lowering of its temperature by expansion ? The reasons hitherto given to explain this coldness are entirely insufficient ; it has been said that the air in high regions, receiving but a small amount of heat reflected by the earth, and itself radiating into celestial space, would lose caloric and thus become colder ; but this explanation is overthrown when we consider that at equal elevations the cold is as great or even greater on elevated plains than on the tops of mountains or in parts of the atmosphere distant from the earth.

gas, or with its density or with its temperature. Experiment has taught us nothing on this subject; it has taught us only that this caloric is developed in greater or less quantity by the compression of elastic fluids.

This preliminary idea having been stated, let us imagine an elastic fluid—atmospheric air, for example—enclosed in a cylindrical vessel *abcd* (Fig. 1) furnished with a movable diaphragm or piston *cd*; let us assume also the two bodies *A*, *B* both at constant temperatures, that of *A* being higher than that of *B*, and let us consider the series of operations which follow :

1. Contact of the body *A* with the air contained in the vessel *abcd* or with the wall of this vessel, which wall is supposed to be a good conductor of caloric. By means of this contact the air attains the same temperature as the body *A*; *cd* is the position of the piston.

2. The piston rises gradually until it takes the position *ef*. Contact is always maintained between the air and the body *A*, and the temperature thus remains constant during the rarefaction. The body *A* furnishes the caloric necessary to maintain a constant temperature.

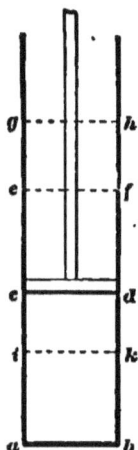

Fig. 1

3. The body *A* is removed and the air is no longer in contact with any body capable of supplying it with caloric; the piston, however, continues to move and passes from the position *ef* to the position *gh*. The air is rarefied without receiving caloric and its temperature falls. Let us suppose that it falls until it becomes equal to that of the body *B*; at this instant the piston ceases to move and occupies the position *gh*.

4. The air is brought in contact with the body *B*; it is compressed by the piston as it returns from the position *gh* to the position *cd*. The air, however, remains at a constant temperature on account of its contact with the body *B*, to which it gives up its caloric.

5. The body B is removed and the compression of the air continued. The temperature of the air, which is now isolated, rises. The compression is continued until the air acquires the temperature of the body A. The piston during this time passes from the position cd to the position ik.

6. The air is again brought in contact with the body A; the piston returns from the position ik to the position ef, and the temperature remains constant.

7. The operation described in No. 3 is repeated, and then the operations 4, 5, 6, 3, 4, 5, 6, 3, 4, · 5, and so on, successively.

In these various operations a pressure is exerted upon the piston by the air contained in the cylinder; the elastic force of this air varies with the changes of volume as well as with the changes of temperature; but we should notice that at equal volumes—that is, for similar positions of the piston—the temperature is higher during the expansions than during the compressions. During the former, therefore, the elastic force of the air is greater, and consequently the quantity of motive power produced by the expansions is greater than that which is consumed in effecting the compressions. Thus there remains an excess of motive power, which we can dispose of for any purpose whatsoever. The air has therefore served as a heat-engine; and it has been used in the most advantageous way possible, for there has been no useless re-establishment of equilibrium in the caloric.

All the operations described above can be carried out in a direct and in a reverse order. Let us suppose that after the sixth step, when the piston is at ef, it is brought back to the position ik, and that, at the same time, the air is kept in contact with the body A; the caloric furnished by this body during the sixth operation returns to its source—that is, to the body A—and the condition of things is the same as at the end of the fifth operation. If now we remove the body A and move the piston from ef to cd, the temperature of the air will fall as many degrees as it rose during the fifth operation and will equal that of the body B. A series of reverse operations to those above described could evidently be carried out; it is only necessary to bring the system into the same initial state and in each operation to carry out an expansion instead of a compression, and conversely.

The result of the first operation was the production of a certain quantity of motive power and the transfer of the caloric from the body *A* to the body *B* ; the result of the reverse operation would be the consumption of the motive power produced and the return of the caloric from the body *B* to the body *A* ; so that the two series of operations in a sense annul or neutralize each other.

The impossibility of making the caloric produce a larger quantity of motive power than that which we obtained in our first series of operations is now easy to prove. It may be demonstrated by an argument similar to that used on page 11. The argument will have even a greater degree of rigor : the air which serves to develop the motive power is brought back, at the end of each cycle of operations, to its original condition, which was, as we noticed, not quite the case with the steam.*

We have chosen atmospheric air as the agency employed to develop the motive power of heat; but it is evident that the same reasoning would hold for any other gaseous substance, and even for all other bodies susceptible of changes of temperature by successive contractions and expansions — that is, for all bodies in Nature, at least, all those which are capable of developing the motive power of heat. Thus we are led to establish this general proposition :

The motive power † *of heat is independent of the agents em-*

---

* We implicitly assume, in our demonstration, that if a body experiences any changes, and returns exactly to its original state, after a certain number of transformations—that is to say, to its original state determined by its density, its temperature, and its mode of aggregation ; we assume, I say, that the body contains the same quantity of heat as it contained at first, or, in other words, that the quantities of heat absorbed and released in its several transformations exactly compensate one another. This fact has never been called in question ; it was at first admitted without consideration and afterwards verified in many cases by experiments with the calorimeter. To deny it would be to overthrow the entire theory of heat, of which it is the foundation. It may be remarked, in passing, that the fundamental principles on which the theory of heat rests should be given the most careful examination. Several experimental facts seem to be almost inexplicable in the actual state of that theory. [The doubts here expressed as to the validity of the assumptions made with respect to the nature of heat developed in Carnot's mind into an actual rejection of those assumptions, and led him to suspect the true nature of heat. See *Life of Carnot.*—Ed.]

† [*It may be well to notice again that Carnot uses* " *motive power* " *as synonymous with the more modern term* " *work.*"]

ployed to develop it ; its quantity is determined solely by the tem-
peratures of the bodies between which, in the final result,* the
transfer of the caloric occurs.

It is understood in this statement that the method used for
developing motive power, whatever it may be, attains the highest
perfection of which it is capable. This condition will be fulfilled, as we remarked above, if there is no change of temperature in the bodies which is not due to a change of volume or,
which amounts to the same thing differently expressed, if the
temperatures of the bodies which come in contact with each
other are never perceptibly different.

Various methods of developing motive power may be adopted,
either by the use of different substances or of the same substance in different states ; for example, by the use of a gas at
two different densities.

This remark leads us naturally to the interesting study of aeriform fluids, a study which will conduct us to new results con-
cerning the motive power of heat, and will give us the means
of verifying in some particular cases the fundamental proposition stated above.†

It can easily be seen that our demonstration will be simplified
if we suppose the temperatures of the bodies *A* and *B* to be very
slightly different. Then the movements of the piston will
be very small during operations 3 and 5, and these operations
may be suppressed without perceptible influence on the development of motive power. That is, a very small change of
volume ought to be sufficient to produce a very small change
of temperature, and this change of volume is negligible compared with that of operations 4 and 6, which are unrestricted
in extent.

If we suppress operations 3 and 5 in the series above described, it is reduced to the following :

1. Contact of the gas contained in *abcd* (Fig. 2) with the body
A, and passage of the piston from *cd* to *ef ;*

2. Removal of the body *A*, contact of the gas enclosed in *abef*
with the body *B*, and return of the piston from *ef* to *cd ;*

3. Removal of the body *B*, contact of the gas with the body

---

* [*That is, upon the completion of a cycle of operations.*]

† We shall suppose in what follows that the reader is familiar with the
latest progress of modern physics in the departments of heat and gases.

A, and passage of the piston from *cd* to *ef*—that is to say, a repetition of the first operation, and so on.

The motive power resulting from the operations 1, 2, 3, taken together, will evidently be the difference between that which is produced by the expansion of the gas while its temperature equals that of the body *A* and that which is consumed to compress the gas while its temperature equals that of the body *B*.

Let us suppose that the operations 1 and 2 are performed with two gases which are chemically different, but which are subjected to the same pressure—for example, that of the atmos-

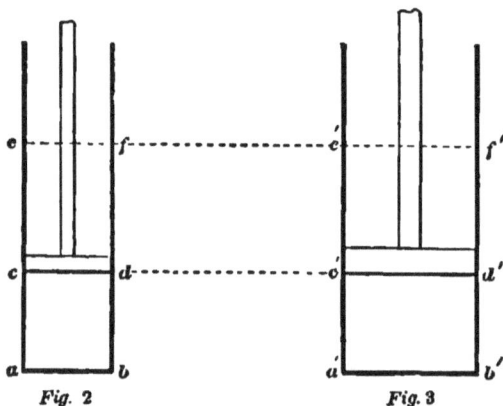

Fig. 2   Fig. 3

phere ; these gases behave in the same circumstances in exactly the same way—that is to say, their expansive forces, originally equal, remain so irrespective of changes of volume and temperature, provided that these changes are the same in both. This is an evident consequence of the laws of Mariotte and of MM. Gay-Lussac and Dalton, which laws are common to all elastic fluids, and in virtue of which the same relations exist in all these fluids between the volume, expansive force, and temperature. Since two different gases, taken at the same temperature and under the same pressure, should behave alike under the same circumstances, they should produce equal quantities of motive power when subjected to the operations above described. Now this implies, according to the fundamental proposition which we have established, that two equal quantities of caloric are employed in these operations—that is, that the quantity of caloric transferred from the body *A* to the body

B is the same whichever of the two gases is used in the operations. The quantity of caloric transferred from the body A to the body B is evidently that which is absorbed by the gas in the increase of its volume, or that which it afterwards emits during compression. We are thus led to lay down the following proposition :

When a gas passes without change of temperature from one definite volume and pressure to another, the quantity of caloric absorbed or emitted is always the same, irrespective of the nature of the gas chosen as the subject of the experiment.

For example, consider 1 litre of air at the temperature of 100 degrees and under the pressure of 1 atmosphere; if the volume of this air is doubled, a certain quantity of heat must be supplied to it in order to maintain it at the temperature of 100 degrees. This quantity will be exactly the same if, instead of performing the operation with air, we use carbonic acid gas, nitrogen, hydrogen, vapor of water, or of alcohol—that is, if we double the volume of 1 litre of any one of these gases at the temperature of 100 degrees and under atmospheric pressure.

The same thing would be true, in the reverse sense, if the volume of the gas, instead of being doubled, were reduced one-half by compression.

The quantity of heat absorbed or set free by elastic fluids during their changes of volume has never been measured by direct experiment. Such an experiment would doubtless present great difficulties, but we have one result which for our purposes is nearly equivalent to it; this result has been furnished by the theory of sound, and may be received with confidence because of the rigor of the demonstration by which it has been established. It may be described as follows :

Atmospheric air will rise in temperature 1 degree centigrade when its volume is reduced by $\frac{1}{116}$ by sudden compression.*

The experiments on the velocity of sound were made in air under a pressure of 760 millimetres of mercury and at the temperature of 6 degrees; and it is only in these circumstances that Poisson's statement is applicable. We shall, however, for the

---

* M. Poisson, to whom we owe this statement, has shown that it agrees very well with the results of an experiment by MM. Clément and Desormes on the behavior of air entering into a vacuum or rather into slightly rarefied air. It agrees also, very nearly, with a result obtained by MM. Gay-Lussac and Welter. (See note, p. 82.)

sake of convenience, consider it to hold at a temperature of 0 degrees, which is only slightly different.

Air compressed by $\frac{1}{116}$ and so raised in temperature 1 degree differs from air heated directly by the same amount only in its density. If we call the original volume V, the compression by $\frac{1}{116}$ reduces it to $V - \frac{1}{116} V$. Direct heating under constant pressure, according to the law of M. Gay-Lussac, should increase the volume of the air by $\frac{1}{267}$ of that which it would have at 0 degrees; thus the volume of the air is in one process reduced to $V - \frac{1}{116} V$, and in the other increased to $V + \frac{1}{267} V$. The difference between the quantities of heat present in the air in the two cases is evidently the quantity used to raise its temperature directly by 1 degree ; thus the quantity of heat absorbed by the air in passing from the volume $V - \frac{1}{116} V$ to the volume $V + \frac{1}{267} V$ is equal to that which is necessary to raise its temperature 1 degree.

Let us now suppose that, instead of heating the air while subjected to a constant pressure and able to expand freely, we enclose it in an envelope not capable of expansion, and then raise its temperature 1 degree. The air thus heated 1 degree differs from air compressed by $\frac{1}{116}$, by having its volume larger by $\frac{1}{116}$. Thus, then, the quantity of heat which the air gives up by a reduction of its volume by $\frac{1}{116}$ is equal to that which is required to raise its temperature 1 degree at constant volume. As the differences, $V - \frac{1}{116} V$, V, and $V + \frac{1}{267} V$, are small in comparison with the volumes themselves, we may consider the quantities of heat absorbed by the air in passing from the first of these volumes to the second, and from the first to the third, as sensibly proportional to the changes of volume. We thus obtain the following relation :

The quantity of heat required to raise the temperature of air under constant pressure 1 degree is to the quantity required to raise it 1 degree at constant volume in the ratio of the numbers
$$\frac{1}{116} + \frac{1}{267} \text{ to } \frac{1}{116},$$
or, multiplying both terms by 116.267, in the ratio of the numbers 267+116 to 267.

This is the ratio between the capacity for heat of air under constant pressure and its capacity at constant volume. If the first of these two capacities is expressed by unity the other will be expressed by the number $\dfrac{267}{267 + 116}$, or, approximately, 0.700.

Their difference—1 — 0.700 or 0.300—will evidently express the quantity of heat which will occasion the increase of volume of the air when its temperature is raised 1 degree under constant pressure.

From the law of MM. Gay-Lussac and Dalton this increase of volume will be the same for all other gases ; from the theorem demonstrated on page 23 the heat absorbed by equal increments of volume is the same for all elastic fluids ; we are thus led to establish the following proposition :

The difference between the specific heat under constant pressure and the specific heat at constant volume is the same for all gases.

It must be noticed here that all the gases are assumed to be taken at the same pressure—for example, the pressure of the atmosphere—and also that the specific heats are measured in terms of the volumes.

Nothing is now easier than to construct a table of the specific heats of gases at constant volume with the aid of our knowledge of their specific heats under constant pressure. We present this table, the first column of which contains the results of direct experiments by MM. Delaroche and Bérard on the specific heat of gases under atmospheric pressure. The second column contains the numbers in the first diminished by 0.300.

TABLE OF THE SPECIFIC HEAT OF GASES

GASES	SPECIFIC HEAT UNDER CONSTANT PRESSURE	SPECIFIC HEAT AT CONSTANT VOLUME
Atmospheric air	1.000	0.700
Hydrogen	0.903	0.603
Carbonic acid..............	1.258	0.958
Oxygen	0.976	0.676
Nitrogen..................	1.000	0.700
Nitrous oxide..............	1.350	1.050
Olefiant gas	1.553	1.253
Carbonic oxide	1.034	0.734

The numbers in the two columns are referred to the same unit, to the specific heat of atmospheric air under constant pressure.

The difference between the corresponding numbers in the two columns being constant, the ratio between them should be variable; thus the ratio between the specific heats of gases under constant pressure and at constant volume varies for the different gases.

We have seen that the temperature of the air when it undergoes a sudden compression of $\frac{1}{116}$ of its volume rises 1 degree. That of other gases should also rise when they are similarly compressed. The temperature should rise, not equally for all, but in the inverse ratio of their specific heats at constant volume. In fact, the reduction of volume being, by hypothesis, always the same, the quantity of heat due to this reduction should also be always the same, and consequently should cause a rise of temperature depending only on the specific heat of the gas after its compression, and evidently in an inverse ratio to that specific heat. It is therefore easy to construct the table of elevations of temperature of the different gases for a compression of $\frac{1}{116}$.

TABLE OF THE ELEVATION OF THE TEMPERATURE OF GASES DUE TO COMPRESSION

GASES	ELEVATION OF TEMPERATURE FOR A REDUCTION OF VOLUME OF $\frac{1}{116}$
	°
Atmospheric air	1.000
Hydrogen	1.160
Carbonic acid	0.730
Oxygen	1.035
Nitrogen	1.000
Nitrous oxide	0.667
Olefiant gas	0.558
Carbonic oxide	0.955

A second compression of $\frac{1}{116}$ of the new volume would, as we shall soon see, again raise the temperature of these gases by an amount nearly equal to the first; but this would not be the case for a third, a fourth, or a hundredth compression of the same sort. The capacity of gases for heat changes with their volume; it is quite possible that it changes also with their temperature.[*]

[*] [*It was found by Regnault* (Mém. de l'Académie, *xxri., p. 53*) *that the specific heat of the "permanent" gases is independent of pressure and temperature.*]

We shall now deduce from the general proposition presented on page 20 a second theorem which will be the complement of that which has just been demonstrated.

Let us suppose that the gas contained in the cylinder *abcd* (Fig. 2) is transferred to the receptacle *a'b'c'd'* (Fig. 3), which is of equal height, but which has a different and larger base ; the gas will increase in volume and diminish in density and elastic force in the inverse ratio of the two volumes *abcd, a'b'c'd'*. The total pressure exerted on each piston, *cd, c'd'*, will be the same, for the surfaces of these pistons are in the direct ratio of the volume.

Let us suppose that the operations described on page 21 as performed on the gas contained in *abcd* are performed on the gas in *a'b'c'd'*—that is, let us suppose that the piston *c'd'* is given displacements equal in amplitude to those given the piston *cd*, and that it occupies successively the positions *c'd'* corresponding to *cd*, and *e'f'* corresponding to *ef*. At the same time let us subject the gas, by means of the two bodies *A, B*, to the same variations of temperature as those to which it was subjected when enclosed in *abcd* ; the total force exerted on the piston will be the same in both cases at corresponding instants. This results immediately from Mariotte's law * ; in fact, the densities of the two gases are always in the same ratio for similar positions of the pistons, and, their temperatures being always equal, the total pressures exerted on the pistons are always in the same ratio. If this ratio is at any time that of equality, the pressures will be always equal.

Further, as the movements of the two pistons have equal amplitudes, the motive power they both produce will evidently be the same, from which we may conclude, from the proposition

---

* Mariotte's law, upon which our demonstration is based, is one of the best-established physical laws. It has served as a foundation for several theories verified by experiment, and which verify in their turn the laws on which they rest. We may also cite, as an important verification of Mariotte's law and also of the law of MM. Gay-Lussac and Dalton for a large range of temperature, the experiments of MM. Dulong and Petit. (See *Annales de Chimie et de Physique*, Feb., 1818, vol. vii., p. 122.) We may also cite the still more recent experiments of Davy and Faraday.

The theorems here deduced would perhaps not be exact if applied outside of certain limits either of density or of temperature. They should only be taken as true within the limits within which the laws of Mariotte, Gay-Lussac, and Dalton are themselves established.

on page 20, that the quantities of heat used by each are equal—that is to say, that the same quantity of heat passes from A to B in each case.

The heat taken from the body A and given to the body B is nothing other than the heat absorbed by the expansion of the gas and afterwards set free by compression. We are thus led to establish the following theorem :

When the volume of an elastic fluid changes, without change of temperature, from U to V, and the volume of a quantity of the same gas, equal in weight and at the same temperature, changes from U' to V', the quantities of heat absorbed or set free from each will be equal when the ratio of U' to V' is equal to that of U to V.

This theorem may be stated in another form, as follows :

When a gas changes in volume without change of temperature the quantities of heat which it absorbs or gives up are in arithmetical progression when the increments or reductions of volume are in geometrical progression.

When we compress one litre of air maintained at a temperature of 10 degrees and reduce its volume to $\frac{1}{2}$ a litre, it gives out a certain quantity of heat. This quantity will be always the same if we further reduce the volume from $\frac{1}{2}$ to $\frac{1}{4}$, from $\frac{1}{4}$ to $\frac{1}{8}$, and so on.

If, instead of compressing the air, we allow it to expand to 2 litres, 4 litres, 8 litres, etc., successively, we must supply it with equal quantities of heat in order to keep its temperature constant.

This easily explains why the temperature of air rises when it is suddenly compressed. We know that this temperature is sufficient to ignite tinder and even to cause the air to become luminous. If we assume for the time being the specific heat of air as constant, in spite of changes of volume and temperature, the temperature will increase in arithmetical progression as the volume is diminished in geometrical progression. Starting with this as given, and admitting that an elevation of temperature of 1 degree corresponds to a compression of $\frac{1}{116}$, it is easy to conclude that when air is reduced to $\frac{1}{14}$ of its original volume its temperature should rise about 300 degrees, which is enough to ignite tinder.*

* When the volume is reduced by $\frac{1}{116}$—that is, when it becomes $\frac{116}{117}$ of that which it was at first—the temperature rises 1 degree.

The elevation of temperature would evidently be still greater if the capacity of the air for heat were to become less as its volume diminishes; now this is probable, and seems to be confirmed by the results of the experiments of MM. Delaroche and Bérard on the specific heat of air taken at different densities. (See the Memoir published in the *Annales de Chimie et de Physique*, vol. lxxxv., pp. 72, 224.)

The two theorems given on pages 23 and 28 are sufficient for the comparison of the quantities of heat absorbed or released in the changes of volume of elastic fluids, whatever may be the density and chemical nature of these fluids, provided always that they are taken and maintained at a certain invariable temperature ; but these theorems do not give us any means of comparing quantities of heat absorbed or released by elastic fluids whose volumes are changed at different temperatures. Thus we do not know the relation between the heat released by 1 litre of air reduced in volume one-half when its temperature is kept at zero and the heat released by the same litre of air reduced in volume one-half when its temperature is kept at 100 degrees. The knowledge of this relation is connected with the knowledge of the specific heat of the gases at different degrees of temperature, and on other data which Physics, in its present state, cannot furnish.

The second of our theorems affords a means of knowing by what law the specific heat of gases varies with their density.

Let us suppose that the operations described on page 21, instead of being performed with two bodies, *A, B,* whose temper-

A new reduction of $\frac{1}{116}$ brings the volume to $(\frac{115}{116})^2$, and the temperature should rise another degree.

After x such reductions the volume is $(\frac{115}{116})^x$, and the temperature should be higher by x degrees.

If we set $(\frac{115}{116})^x = \frac{1}{2}$, and take the logarithms of both sides, we find $x = 300°$ about.

If we set $(\frac{115}{116})^x = \frac{1}{2}$ we find that $x = 80°$, which shows that the temperature of air compressed to one half of its original volume rises 80 degrees.

All this is dependent on the hypothesis that the specific heat of air does not change when the volume diminishes ; but if, for the reasons given on pages 31 and 32, we consider the specific heat of air compressed to one-half its volume as reduced in the ratio of 700 to 616, we must multiply 80 degrees by $\frac{700}{616}$, which brings it to 90 degrees.

ntures differ by an infinitely small quantity, are performed with two bodies whose temperatures differ by a finite quantity, say by 1°.

In a complete cycle of operations the body A furnishes to the elastic fluid a certain quantity of heat which may be divided into two portions: 1, the quantity required to keep the temperature of the fluid constant during expansion; 2, that required to change the temperature of the fluid from that of the body B to that of the body A, after the fluid has been restored to its original volume and is put in contact with the body A. Let us call the first of these quantities a and the second b. The total caloric furnished by the body A will be expressed by $a + b$.

The caloric transmitted by the fluid to the body B may also be divided into two parts; one of which, b', is due to the cooling of the gas by the body B, the other, a', is that released by the gas during the reduction of its volume. The sum of these two quantities is $a' + b'$; this should be equal to $a + b$, for after a complete cycle of operations the gas returns exactly to its original state.* It must have given up all the caloric with which it had at first been supplied. We then have

$$a + b = a' + b',$$

or, $$a - a' = b - b'.$$

Now, from the theorem given on page 28, the quantities a and a' are independent of the density of the gas, always provided that the quantity of the gas by weight remains the same and that the variations of volume are proportional to the original volume. The difference $a - a'$ should satisfy the same conditions, and consequently also the difference $b' - b$, which is equal to it. But b' is the caloric necessary to raise the temperature of the gas contained in $abcd$ one degree (Fig. 2); b' is the caloric released by the gas when it is enclosed in $abef$, and its temperature falls one degree. These quantities can serve as a measure of the specific heats. We are thus led to establish the following proposition:

* [*The use here made of the caloric theory vitiates the demonstration and leads to erroneous conclusions.*]

The change made in the specific heat of a gas in consequence of a change of volume depends only upon the relation between the original volume and that which results from the change—that is to say, the difference between the specific heats does not depend on the absolute magnitudes of the volumes but on their ratio.

This proposition may be stated in another way, namely :

When the volume of a gas increases in geometrical progression its specific heat increases in arithmetical progression.

Thus, if a is the specific heat of air taken at a given density, and $a + h$ its specific heat when its density is one-half this, its specific heat will be $a + 2h$ when its density is one-quarter this, $a + 3h$ when its density is one-eighth this, and so on.

The specific heats are here referred to weight. They are supposed to be taken at constant volume ; but, as we shall see, they would follow the same law if they were taken under constant pressure.

In fact, to what cause is due the difference between the specific heats taken at constant volume and under constant pressure ? It is due to the caloric required in the latter case to produce the increase of volume. Now, by Mariotte's law, the increase of volume of a gas, for a given change of temperature, should be a definite fraction of the original volume, which fraction is independent of the pressure. From the theorem given on page 28, if the ratio between the original volume and the changed volume is given, the heat required to produce the increase of volume is determined thereby. It depends only on this ratio and on the quantity of the gas by weight. We must then conclude that : *The difference between the specific heat under constant pressure and that at constant volume is always the same, whatever the density of the gas, provided that the quantity of the gas by weight remains the same.* These specific heats both increase as the density of the gas diminishes, but their difference does not change.* Since the difference between the two capaci-

---

* MM. Gay-Lussac and Welter have found by direct experiments, cited in the *Mécanique Céleste* and in the *Annales de Chimie et de Physique*, July, 1822, page 267, that the ratio between the specific heat under constant pressure and that at constant volume varies very little with the density of the gas. From what we have just seen, it is the difference and not the ratio that should remain constant. However, as the specific heats of gases, for a given weight, vary very little with their density, it is clear that the ratio also will experience only very small changes.

tics for heat is constant, when one increases in arithmetical progression the other will increase in a similar progression; thus our law applies to specific heats taken under constant pressure.

We have tacitly supposed that the specific heat increases with the volume. This increase is shown in the experiments of MM. Delaroche and Bérard; these physicists have found that the specific heat of air under the pressure of 1 meter of mercury is 0.967 (see the memoir already referred to), taking as the unit the specific heat of the same weight of air under the pressure of 0.760 meter. From the law followed by the specific heats with respect to pressure, observations made of them in two particular cases permit us to calculate them in all possible cases; thus, by using the result of the experiment of MM. Delaroche and Bérard, which has just been cited, we have constructed the following table of the specific heats of air under various pressures:

PRESSURE IN ATMOSPHERES	SPECIFIC HEAT, THAT OF AIR UNDER ATMOSPHERIC PRESSURE BEING 1.	PRESSURE IN ATMOSPHERES	SPECIFIC HEAT, THAT OF AIR UNDER ATMOSPHERIC PRESSURE BEING 1.
$\frac{1}{1024}$	1.840	1	1.000
$\frac{1}{512}$	1.756	2	0.916
$\frac{1}{256}$	1.672	4	0.832
$\frac{1}{128}$	1.588	8	0.748
$\frac{1}{64}$	1.504	16	0.664
$\frac{1}{32}$	1.420	32	0.580
$\frac{1}{16}$	1.336	64	0.496
$\frac{1}{8}$	1.252	128	0.412
$\frac{1}{4}$	1.165	256	0.328
$\frac{1}{2}$	1.084	512	0.244
1	1.000	1024	0.160

From the experiments of MM. Gay-Lussac and Welter, the ratio of the specific heat of atmospheric air under constant pressure to that at constant volume is 1.3748, a number which is nearly constant for all pressures and for all temperatures. In the previous discussion we have been led, by other considerations, to the number $\frac{267+116}{267} = 1.44$, which differs from this by $\frac{1}{16}$, and we have used this number to construct a table of the specific heats of gases at constant volume. Neither this table nor the table given on page 32 should be considered as accurate. They are intended mainly to set forth the laws followed by the specific heats of aeriform fluids.

The numbers in the first column are in geometrical progression, while those in the second are in arithmetical progression. We have carried the table out to extreme compressions and rarefactions. It is to be supposed that air, before attaining a density 1024 times its ordinary density—that is, before becoming more dense than water—would be liquefied. The specific heats vanish, and even become negative if we prolong the table beyond the last number given. It seems probable that the numbers in the second column decrease too rapidly. The experiments on which we based our calculation were made within too narrow limits to enable us to expect great exactness in the numbers obtained, especially in the extreme values.

Since, on the one hand, we know the law by which heat is evolved by the compression of a gas, and, on the other, the law by which the specific heat varies with the volume, it will be easy for us to calculate the increase of temperature of a gas compressed without loss of caloric.* In fact, the compression can be considered as consisting of two successive operations : 1, compression at a constant temperature, and, 2, restoration of the caloric emitted. In the second operation the temperature will rise in the inverse ratio to the specific heat which the gas acquires by the reduction of its volume. We can determine the specific heat by means of the law above demonstrated. From the theorem on page 28 the heat set free by compression should be represented by an expression of the form

$$s = A + B \log v,$$

s being the heat, v the volume of the gas after compression, A and B arbitrary constants dependent on the original volume of the gas, on its pressure, and on the units which are chosen.

The specific heat, which varies with the volume in accordance with the law just demonstrated, should be represented by an expression of the form,

$$z = A' + B' \log v,$$

A' and B' being arbitrary constants different from A and B.

The increase of temperature which the gas receives by compression is proportional to the ratio $\frac{s}{z}$, or to the ratio $\frac{A + B \log v}{A' + B' \log v}$.

* [*This demonstration is erroneous in that it assumes the materiality of heat, and also the change of specific heat with volume. The conclusions are invalid.*]

It may be represented by this ratio itself; thus, if we represent it by t, we shall have
$$t = \frac{A + B \log v}{A' + B' \log v}.$$
If the original volume of the gas is 1 and the original temperature zero, we shall have at the same time $t = 0$, $\log v = 0$, and hence $A = 0$; t will then express not only the increase of temperature, but the temperature itself above the thermometric zero.

We must not think that we can apply the formula just given to very large changes in the volume of the gas. We have taken the rise of temperature to be in the inverse ratio to the specific heat, which implies that the specific heat is constant at all temperatures. Large changes of volume in the gas occasion large changes of temperature, and there is no evidence that the specific heat is constant at different temperatures, especially when these temperatures are widely separated from each other. This constancy of specific heat is only an hypothesis assumed in the case of gases from analogy, and verified fairly well for solids and liquids within a certain range of the thermometric scale, but which the experiments of MM. Dulong and Petit have shown to be inexact when extended to temperatures much above 100 degrees.[*]

---

[*] We see no reason to assume *a priori* the constancy of the specific heat of bodies at various temperatures — that is to say, to assume that equal quantities of heat will produce equal increments in the temperature of a body, even when neither the state nor the density of the body is changed; as, for example, in the case of an elastic fluid enclosed in a rigid envelope. Direct experiments on solid and liquid bodies have proved that between zero and 100 degrees equal increments of heat produce nearly equal increments of temperature; but the more recent experiments of MM. Dulong and Petit (see *Annales de Chimie et de Physique*, February, March, and April, 1818) have shown that this relation does not hold for temperatures much over 100 degrees, whether they are measured by the mercury thermometer or by the air thermometer.

Not only do the specific heats not remain the same at different temperatures, but the ratios between them do not remain the same; so that no thermometric scale can establish the constancy of all specific heats at the same time. It would be interesting to examine whether the same irregularities would obtain in gaseous substances, but the necessary experiments present almost insurmountable difficulties.

It seems probable that the irregularities of the specific heats of solid bodies may be attributed to latent heat, employed in producing a commencement of fusion, a softening which in many cases becomes perceptible in these bodies long before complete fusion occurs. We can support this opinion by the following observation: From the experiments of MM. Dulong and Petit, the increase of specific heat with the temperature is more rapid

THE SECOND LAW OF THERMODYNAMICS

According to a law due to MM. Clément and Desormes,* established by direct experiment, water vapor, under whatever pressure it is formed, always contains the same quantity of heat in equal weights ; this amounts to saying that the vapor when compressed or expanded without loss of heat is always in such a condition as to saturate the space which it occupies, if it is originally in this condition. Water vapor in this condition can thus be considered a permanent gas, and should follow all the laws of gases. Consequently, the formula

$$t = \frac{A + B \log v}{A' + B' \log v}$$

should be applicable and should agree with the table of tensions constructed by M. Dalton from his direct experiments.

We can, in fact, satisfy ourselves that our formula, with a suitable determination of the arbitrary constants, represents very approximately the results of experiment. The unimportant discrepancies which we find in it are no more than may reasonably be attributed to errors of observation.†

with solids than with liquids, though the latter have a larger dilatability. If the cause which we have proposed to account for this irregularity is the real one, it would disappear entirely with gases.

* [*It has been shown by Rankine and Clausius that this law does not hold.*]
† To determine the arbitrary constants A, B, A', B', from data taken from M. Dalton's table, we must begin by calculating the volume of the vapor by means of its pressure and temperature, the quantity of the vapor by weight being always constant. This is made easy by the laws of Mariotte and Gay-Lussac. The volume will be given by the equation

$$v = c\,\frac{267+t}{p},$$

in which v is the volume, t the temperature, p the pressure, and c a constant quantity which depends on the weight of the vapor and the units chosen.

The following is the table of the volumes occupied by a gram of vapor formed at various temperatures and consequently under various pressures :

t OR CENTIGRADE DEGREES	*p* OR THE TENSION OF THE VAPOR EXPRESSED IN MILLIMETRES OF MERCURY	*v* OR THE VOLUME OF A GRAM OF VAPOR EXPRESSED IN LITRES
Deg.	*Mm.*	*Lit.*
0	5.060	185.0
20	17.32	58.2
40	58.00	20.4
60	144.6	7.96
80	352.1	3.47
100	760.0	1.70

The first two columns in this table are taken from the *Traité de Physique*

We shall now return to our principal subject, the motive power of heat, from which we have already digressed too far.

We have shown that the quantity of motive power developed by the transfer of caloric from one body to another depends essentially on the temperatures of the two bodies, but we have not discussed the relation between these temperatures and the quantities of motive power produced. It would seem at first natural enough to suppose that for equal differences of temperature the quantities of motive power produced are equal— that is, for example, that a given quantity of caloric passing from a body, A, kept at 100 degrees, to a body, B, kept at 50 degrees would develop a quantity of motive power equal to that which would be developed by the transfer of the same caloric from a body, B, kept at 50 degrees to a body, C, kept at zero. Such a law would indeed be a very remarkable one, but we do not see sufficient reason to admit it *a priori*. We shall examine this question by a rigorous method.

Let us suppose that the operations described on page 21 are performed successively on two quantities of atmospheric air equal in weight and volume but taken at different temperatures, and let us suppose also that the differences of temperature be-

---

of M. Biot (vol. i., pp. 272 and 531). The third is calculated by means of the above formula, and from the experimental fact that the vapor of water under atmospheric pressure occupies a volume 1700 times as great as that which it occupies when in the liquid state.

By using three numbers from the first column and the corresponding numbers from the third, we can easily determine the constants of our equation

$$t = \frac{A + B \log v}{A' + B' \log v}.$$

We shall not enter into the details of the calculation necessary to determine these quantities; it will be enough for us to say that the following values,

$$A = 2268, \qquad A' = 19.64,$$
$$B = -1000, \qquad B' = 3.30,$$

satisfy sufficiently well the prescribed conditions, so that the equation

$$t = \frac{2268 - 1000 \log v}{19.64 + 3.30 \log v}$$

expresses very approximately the relation existing between the volume of the vapor and its temperature.

It is to be noticed that the quantity B' is positive and very small, which tends to confirm the proposition that the specific heat of an elastic fluid increases with the volume, but at a very slow rate.

tween the bodies A and B are the same in both cases; thus, for example, the temperatures of these bodies will be in one case $100°$ and $100° - h$ (h being infinitely small), and in the other, $1°$ and $1° - h$. The quantity of motive power produced is in each case the difference between that which the gas furnishes by its expansion and that which must be used to restore it to its original volume. Now this difference is here the same in both cases, as we may satisfy ourselves by a simple argument, which we do not think it necessary to give in full; so that the motive power produced is the same. Let us now compare the quantities of heat used in the two cases. In the first case the quantity used is that which the body A imparts to the air in order to keep it at a temperature of 100 degrees during its expansion; in the second, it is that which the same body imparts to it to maintain its temperature at 1 degree during an exactly similar change of volume. If these two quantities were equal it is evident that the law which we have assumed would follow. But there is nothing to prove that it is so; we proceed to prove that these quantities of heat are unequal.

The air which we first supposed to occupy the space $abcd$ (Fig. 2) and to be at a temperature of 1 degree, may be made to occupy the space $abef$, and to acquire the temperature of 100 degrees by two different methods:

1. It may first be heated without change of volume, and then expanded while its temperature is kept constant.

2. It may first be expanded while its temperature is kept constant, and then heated when it has acquired its new volume.

Let a and b be the quantities of heat used successively in the first of the two operations, and b' and a' the quantities used in the second; as the final result of these two operations is the same, the quantities of heat used in each should be equal; we then obtain

$$a + b = a' + b',$$

from which we have

$$a' - a = b - b'.$$

We represent by a' the quantity of heat necessary to raise the temperature of the gas from 1 to 100 degrees when it occupies the volume $abef$, and by a the quantity of heat necessary to raise the temperature of the gas from 1 to 100 degrees when it occupies the volume $abcd$.

The density of the air is less in the first case than in the second, and from the experiments of MM. Delaroche and Bérard, already cited on page 32, its capacity for heat should be a little greater.

As the quantity a' is greater than the quantity a, b should be greater than b', consequently, stating the proposition generally, we may say that :

The quantity of heat due to the change of volume of a gas becomes greater as the temperature is raised.

Thus, for example, more caloric is required to maintain at 100 degrees the temperature of a certain quantity of air whose volume is doubled than to maintain at 1 degree the temperature of the same quantity of air during a similar expansion.

These unequal quantities of heat will, however, as we have seen, produce equal quantities of motive power for equal descents of caloric occurring at different heights on the thermometric scale ; from which we may draw the following conclusion :

*The descent of caloric produces more motive power at lower degrees of temperature than at higher.**

Thus a given quantity of heat will develop more motive power in passing from a body whose temperature is kept at 1 degree to another whose temperature is kept at zero than if the temperatures of these two bodies had been 101 and 100 respectively. It must be said that the difference should be very small ; it would be zero if the capacity of air for heat remained constant in spite of changes of density. According to the experiments of MM. Delaroche and Bérard, this capacity varies very little, so little, indeed, that the differences noticed might strictly be attributed to errors of observation or to some circumstances which were not taken into account.

It would be out of the question for us, with the experimental data at our command, to determine rigorously the law by which

---

* [*The preceding demonstration is erroneous in consequence of the assumption of the materiality of heat. The conclusion is in form correct, but only because of the erroneous use of a variable specific heat of air. If this be considered constant, as Carnot points out, the efficiency should be, on his principles, the same at all temperatures. The ratio of the motive power produced to the heat used should be equal to the difference of temperature multiplied by a constant, the " Carnot's function." As we now know, this function is not constant, but is the reciprocal of the absolute temperature of the source of heat.*]

the motive power of heat varies at different degrees of the thermometric scale. It is connected with the law of the variations of the specific heat of gases at different temperatures, which has not been determined with sufficient exactness.* We

---

* If we admit that the specific heat of a gas is constant when its volume does not change, but only its temperature varies, analysis would lead us to a relation between the motive power and the thermometric degree. We shall now examine the way in which this may be done; it will also give us an opportunity of showing how some of the propositions formerly established should be stated in algebraic form.

Let r be the quantity of motive power produced by the expansion of a given quantity of air changing from the volume 1 litre to the volume v litres at constant temperature. If v increases by the infinitely small quantity dv, r will increase by the quantity dr, which, from the nature of motive power, will be equal to the increase of volume dv multiplied by the expansive force which the elastic fluid then has. If p represents the expansive force, we shall have the equation

(1) $$dr = p\,dv.$$

Let us suppose the constant temperature at which the expansion occurs to be equal to t degrees centigrade. Representing by q the elastic force of the air at the same temperature, t, occupying the volume of 1 litre, we shall have from Mariotte's law

$$\frac{v}{1} = \frac{q}{p}, \text{ from which } p = \frac{q}{v}.$$

Now if P is the elastic force of the same air always occupying the volume 1, but at the temperature zero, we shall have from M. Gay-Lussac's law

$$q = P + P\frac{t}{267} = \frac{P}{267}(267 + t)\ ;$$

from which

$$\frac{q}{v} = p = \frac{P}{267}\frac{267 + t}{v}.$$

If, for the sake of brevity, we represent by N the quantity $\frac{P}{267}$, the equation will become

$$p = N\frac{t + 267}{v}\ ;$$

by using which we have, from equation (1),

$$dr = N\frac{t + 267}{v}\,dv.$$

Considering t constant, and taking the integrals of the two terms, we obtain

$$r = N(t + 267)\log v + C.$$

If we suppose that $r = 0$ when $v = 1$, we shall have $C = 0$, from which

(2) $$r = N(t + 267)\log v.$$

This is the motive power produced by the expansion of the air at the temperature t, whose volume has changed from 1 to v. If instead of working

shall now endeavor to determine definitively the motive power of heat, and in order to verify our fundamental proposition—

at the temperature t we work in exactly the same way at the temperature $t+dt$, the power developed will be

$$r + \delta r = N(t + dt + 267) \log v.$$

Subtracting equation (2) we obtain

(3) $$\delta r = N \log v dt.$$

Let e be the quantity of heat used to keep the temperature of the gas constant during its expansion. From the discussion on page 21 δr will be the power developed by the descent of the quantity of heat e from the degree $t+dt$ to the degree t. Let u represent the motive power developed by the descent of a unit of heat from t degrees to zero; since from the general principle established on page 21 this quantity u should depend only on t, it may be represented by the function Ft, from which $u = Ft$.

When t increases and becomes $t+dt$, u becomes $u + du$, from which

$$u + du = F(t + dt).$$

Subtracting the preceding equation we have

$$du = F(t + dt) - Ft = F'tdt.$$

This is evidently the quantity of motive power produced by the descent of a unit quantity of heat from the degree $t + dt$ to the degree t.

If the quantity of heat, instead of being a unit, had been e, the motive power produced would have been

(4) $$edu = eF'tdt.$$

But edu is the same as δr, both being the power developed by the descent of the quantity of heat e from the degree $t+dt$ to the degree t; consequently,

$$edu = \delta r,$$

and, from equations (3) and (4),

$$eF'tdt = N \log v dt;$$

or, dividing by $F'tdt$, and representing by T the fraction $\dfrac{N}{F't}$, which is a function of t only, we have

$$e = \frac{N}{F't} \log v = T \log v.$$

The equation $\qquad e = T \log v$

is the analytical expression of the law stated on page 28; it is the same for all gases, since the laws we have used are common to all.

If we represent by s the quantity of heat required to change the volume of the air with which we are working from 1 to v, and the temperature from zero to t, the difference between s and e will be the quantity of heat required to change the temperature of the air, while its volume remains 1, from zero to t. This quantity depends on t only. It will be some function of t, and we shall have, if we call it U,

$$s = e + U = T \log v + U.$$

If we differentiate this equation with respect to t only and represent by T' and U' the differential coefficients of T and U, it will become

(5) $$\frac{ds}{dt} = T' \log v + U';$$

40

that is, to show that the quantity of motive power produced is really independent of the agent used—we shall choose sev-

$\frac{ds}{dt}$ is nothing other than the specific heat of the gas at constant volume, and our equation (5) is the analytical expression of the law stated on page 31.

If we suppose the specific heat to be constant at all temperatures—an hypothesis which was discussed on page 34—the quantity $\frac{ds}{dt}$ will be independent of t, and, to satisfy equation (5) for two particular values of v, T'' and U' must also be independent of t; we shall then have $T''=C$, a constant quantity. Multiplying T'' and C by dt and integrating both sides we find

$$T = Ct + C_1 ;$$

but as $T=\frac{N}{F''t}$, we have

$$F't = \frac{N}{T} = \frac{N}{Ct + C_1}.$$

Multiplying both sides by dt and integrating we obtain

$$Ft = \frac{N}{C} \log (Ct + C_1) + C_2 ;$$

or, changing the arbitrary constants, and remembering that Ft is zero when $t = 0°$, we have

(6) $$Ft = A \log \left(1 + \frac{t}{B}\right).$$

The nature of the function Ft is thus determined, and may serve us as a means of calculating the motive power developed by any descent of heat. But this last conclusion is based on the hypothesis of the constancy of the specific heat of a gas whose volume does not change—an hypothesis which experiment has not yet sufficiently verified. Until there are further proofs of its validity equation (6) can only be admitted for a small part of the thermometric scale.

The first term in equation (5) represents, as we have said, the specific heat of the air occupying the volume v. Experiment has taught us that this specific heat varies only slightly in spite of considerable changes of volume, so that the coefficient T' of $\log v$ must be a very small quantity. If we assume that it is zero and multiply the equation $T'=0$ by dt and then integrate, we have

$$T = C, \text{ a constant quantity.}$$

But

$$T = \frac{N}{F't},$$

from which

$$F''t = \frac{N}{T} = \frac{N}{C} = A ;$$

from which we may conclude by a second integration that

$$Ft = At + B.$$

eral such agents — atmospheric air, water vapor, and alcohol vapor.

Let us take first atmospheric air. The operation is effected according to the method indicated on page 21. We make the following hypotheses:

The air is taken under atmospheric pressure; the temperature of the body A is $\frac{1}{1000}$ of a degree above zero and that of the body B is zero. We see that the difference is, as it should be, very small. The increase of the volume of the air in our operation will be $\frac{1}{116} + \frac{1}{267}$ of the original volume; this is a very small increase considered absolutely, but large relatively to the difference of temperature between A and B.

The motive power developed by the two operations described on page 21 taken together will be very nearly proportional to the increase of volume and to the difference between the two pressures exerted by the air when its temperature is $0.001°$ and zero.

According to the law of M. Gay-Lussac, this difference is $\frac{1}{267000}$ of the elastic force of the gas, or very nearly $\frac{1}{267000}$ of the atmospheric pressure.

The pressure of the atmosphere is equal to that of a column of water $10\frac{40}{100}$ meters high; $\frac{1}{267000}$ of this pressure is equal to that of a water column $\frac{1}{267000} \times 10.40$ meters in height.

As for the increase of volume, it is, by hypothesis, $\frac{1}{116} + \frac{1}{267}$ of the original volume—that is, of the volume occupied by 1 kilogram of air at zero, which is equal to 0.77 cubic meters, if we take into account the specific gravity of air; thus the product,

$$\left(\tfrac{1}{116} + \tfrac{1}{267}\right) 0.77 \tfrac{1}{267000} 10.40$$

expresses the motive power developed. This power is here estimated in cubic meters of water raised to the height of 1 meter.

If we carry out the multiplications indicated, we find for the product 0.000000372.

Let us now try to determine the quantity of heat used to obtain this result—that is, the quantity transferred from the body A to the body B. The body A furnishes:

As $Ft = 0$ when $t = 0$, B is zero; thus
$$Ft = At;$$
that is to say, the motive power produced is exactly proportional to the descent of the caloric. This is the analytical expression of the statement made on page 38.

1. The heat required to raise the temperature of 1 kilogram of air from zero to 0.001°.

2. The quantity required to maintain the temperature of the air at 0.001° when it undergoes an expansion of

$$\tfrac{1}{116} + \tfrac{1}{267}.$$

The first of these quantities of heat may be neglected, as it is very small in comparison with the second, which is, from the discussion on page 24, equal to that required to raise the temperature of 1 kilogram of air under atmospheric pressure 1 degree.

The specific heat of air by weight is 0.267 that of water, from the experiments of MM. Delaroche and Bérard on the specific heat of gases. If, then, we take for the unit of heat the quantity required to raise 1 kilogram of water 1 degree, the quantity required to raise 1 kilogram of air 1 degree will be 0.267. Thus the quantity of heat furnished by the body A is

0.267 unit.

This quantity of heat is capable of producing 0.000000372 unit of motive power by its descent from 0.001 to zero.

For a descent one thousand times as great, or of one degree, the motive power will be very nearly one thousand times as great as this, or

0.000372.

Now if, instead of using 0.267 unit of heat, we use 1000 units, the motive power produced will be given by the proportion

$$\tfrac{0.267}{0.000372} = \tfrac{1000}{x}, \text{ from which } x = \tfrac{372}{267} = 1.395 \text{ units.}$$

Thus if 1000 units of heat pass from a body whose temperature is kept at 1 degree to another at zero, they will produce by their action on air 1.395 units of motive power.

We shall compare this result with that which is obtained from the action of heat on water vapor.

Let us suppose that 1 kilogram of water is contained in the cylinder $abcd$ (Fig. 4) between the base ab and the piston cd, and let us assume also the existence of two bodies, A, B, each maintained at a constant temperature, that of A being higher than that of B by a very small quantity. We shall now imagine the following operations :

Fig. 4

1. Contact of the water with the body A, change of the position of the piston from cd to ef, formation of vapor at the temperature of the body A to fill the vacuum made by the increase of the volume. We shall assume the volume $abef$ to be large enough to contain all the water in a state of vapor;

2. Removal of the body A, contact of the vapor with the body B, precipitation of a part of this vapor, decrease of its elastic force, return of the piston from ef to ab, and liquefaction of the rest of the vapor by the effect of the pressure combined with the contact of the body B;

3. Removal of the body B, new contact of the water with the body A, return of the water to the temperature of this body, a repetition of the first operation, and so on.

The quantity of motive power developed in a complete cycle of operations is measured by the product of the volume of the vapor multiplied by the difference between its tensions at the temperatures of the body A and of the body B respectively.

The heat used—that is, that transferred from the body A to the body B—is evidently the quantity which is required to transform the water into vapor, always neglecting the small quantity necessary to restore the water from the temperature of the body B to that of the body A.

Let us suppose that the temperature of the body A is 100 degrees and that of the body B 99 degrees. From M. Dalton's table the difference of these tensions will be 26 millimetres of mercury or 0.36 meter of water. The volume occupied by the vapor is 1700 that of the water, so that, if we use 1 kilogram, it will be 1700 litres or 1.700 cubic meters. Thus the motive power developed is

$$1.700 \times 0.36 = 0.611 \text{ unit}$$

of the sort which we used before.

The quantity of heat used is the quantity required to transform the water into vapor, the water being already at a temperature of 100 degrees. This quantity has been determined by experiment; it has been found equal to 550 degrees, or, speaking with greater precision, to 550 of our units of heat.

Thus 0.611 unit of motive power result from the use of 550 units of heat.

The quantity of motive power produced by 1000 units of heat will be given by the proportion

$$\frac{550}{0.611} = \frac{1000}{x}, \text{ from which } x = \frac{611}{550} = 1.112.$$

Thus 1000 units of heat transferred from a body maintained at 100 degrees to one maintained at 99 degrees will produce 1.112 units of motive power when acting on the water vapor. The number 1.112 differs by nearly $\frac{1}{4}$ from 1.395, which was the number previously found for the motive power developed by 1000 units of heat acting on air; but we must remember that in that case the temperature of the bodies A and B were 1 degree and zero, while in this case they are 100 and 99 degrees respectively. The difference is indeed the same, but the temperatures on the thermometric scale are not the same. In order to obtain an exact comparison it would be necessary to calculate the motive power developed by the vapor formed at 1 degree and condensed at zero, and also to determine the quantity of heat contained in the vapor formed at 1 degree. The law of MM. Clément and Desormes, to which we referred on page 35, gives us this information. The heat used in turning water into vapor (*chaleur constituante*) is always the same at whatever temperature the vaporization occurs. Therefore, since 550 degrees of heat are required to vaporize the water at the temperature of 100 degrees, we must have 550 + 100, or 650 degrees, to vaporize the same weight of water at zero.

By using the data thus obtained, and reasoning in other respects quite in the same way as we did when the water was at 100 degrees, we readily see that 1.290 is the motive power developed by 1000 units of heat acting on water vapor between the temperatures of 1 degree and zero.

This number approaches 1.395 more nearly than the other.

It only differs by $\frac{1}{13}$, which is not outside the limits of probable error, considering the large number of data of different sorts which we have found it necessary to use in making this comparison. Thus our fundamental law is verified in a particular case.*

---

* In a memoir of M. Petit (*Annales de Chimie et de Physique*, July, 1818, page 294) there is a calculation of the motive power of heat applied to air and to water vapor. The results of this calculation are much to the advantage of atmospheric air ; but this is owing to a very inadequate way of considering the action of heat.

We shall now examine the case of heat acting on alcohol vapor. The method used in this case is exactly the same as in the case of water vapor, but the data are different. Pure alcohol boils under ordinary pressure at 78.7° centigrade. According to MM. Delaroche and Bérard, 1 kilogram of this substance absorbs 207 units of heat when transformed into vapor at this same temperature, 78.7°.

The tension of alcohol vapor at 1 degree below its boiling-point is diminished by $\frac{1}{23}$, and is $\frac{1}{25}$ less than atmospheric pressure (this is at least the result of the experiments of M. Bétancour, an account of which was given in the second part of M. Prony's *Architecture Hydraulique*, pages 180, 195).*

We find, by use of these data, that the motive power developed, in acting on 1 kilogram of alcohol at the temperatures 77.7° and 78.7°, would be 0.251 unit.

This results from the use of 207 units of heat. For 1000 units we must set the proportion

$$\frac{207}{0.254} = \frac{1000}{x}, \text{ from which } x = 1.230.$$

This number is a little greater than 1.112, resulting from the use of water vapor at 100 and 99 degrees; but if we assume the water vapor to be employed at 78 and 77 degrees, we find, by the law of MM. Clément and Desormes, 1.212 for the motive power produced by 1000 units of heat. As we see, this number approaches 1.230 very nearly; it only differs from it by $\frac{1}{66}$.

* M. Dalton thought that he had discovered that the vapors of different liquids exhibited equal tensions at temperatures on the thermometric scale equally distant from their boiling-points; this law is, however, not rigorously, but only approximately, correct. The same is true of the law of the ratio of the latent heat of vapors to their densities (see extracts from a memoir of M. C. Despretz, *Annales de Chimie et de Physique*, vol. xvi., p. 105, and vol. xxiv., p. 323). Questions of this kind are closely connected with those relating to the motive power of heat. Davy and Faraday recently tried to recognize the changes of tension of liquefied gases for small changes of temperature, after having made excellent experiments on the liquefaction of gases by the effect of a considerable pressure. They had in view the use of new liquids in the production of motive power (see *Annales de Chimie et de Physique*, January, 1824, p. 80). From the theory given above we can predict that the use of these liquids presents no advantage for the economical use of heat. The advantage could only be realized at the low temperature at which it would be possible to work, and by the use of sources from which, for this reason, it would become possible to extract caloric.

We should have liked to have made other comparisons of this kind—for example, to have calculated the motive power developed by the action of heat on solids and liquids, by the freezing of water, etc.; but in the present state of Physics we are not able to obtain the necessary data.* The fundamental law which we wish to confirm seems, however, to need additional verifications to be put beyond doubt; it is based upon the theory of heat as it is at present established, and, it must be confessed, this does not appear to us to be a very firm foundation. New experiments alone can decide this question; in the mean time we shall occupy ourselves with the application of the theoretical ideas above stated, and shall consider them as correct in the examination of the various means proposed at the present time to realize the motive power of heat.

It has been proposed to develop motive power by the action of heat on solid bodies. The mode of procedure which most naturally presents itself to our minds is to firmly fix a solid body—a metallic bar, for example—by one of its extremities, and to attach the other extremity to a movable part of the machine; then by successive heating and cooling to cause the length of the bar to vary, and thus produce some movement. Let us endeavor to decide if this mode of developing motive power can be advantageous. We have shown that the way to get the best results in the production of motion by the use of heat is to so arrange the operations that all the changes of temperature which occur in the bodies are due to changes of volume. The more nearly this condition is fulfilled the better heat will be utilized. Now, by proceeding in the manner just described, we are far from fulfilling this condition; no change of temperature is here due to a change of volume; but the changes are all due to the contact of bodies differently heated, to the contact of the metallic bar either with the body which furnishes the heat or with the body which absorbs it.

The only means of fulfilling the prescribed condition would be to act on the solid body exactly as we did on the air in the operations described on page 18, but for this we must be able to produce considerable changes of temperature solely by the change of volume of the solid body, if, at least, we desire to

* The data lacking are the expansive force acquired by solids and liquids for a given increase of temperature, and the quantity of heat absorbed or emitted during changes in the volume of these bodies.

use considerable descents of caloric. Now this seems to be impracticable, for several considerations lead us to think that the changes in the temperature of solids or liquids by compression and expansion are quite small.

1. We often observe in engines (in heat-engines particularly) solid parts which are subjected to very considerable forces, sometimes in one sense and sometimes in another, and although those forces are sometimes as great as the nature of the substances employed will permit, the changes in temperature are scarcely perceptible.

2. In the process of striking medals, of rolling plates, or of drawing wires, metals undergo the greatest compressions to which we can subject them by the use of the hardest and most resisting materials. Notwithstanding this the rise in temperature is not great, for if it were, the steel tools which we use in these operations would soon lose their temper.

3. We know that it is necessary to exert a very great force on solids and liquids to produce in them a reduction of volume comparable to that which they undergo by cooling (for example, by a cooling from 100 degrees to zero). Now, cooling requires a greater suppression of caloric than would be required by a simple reduction of volume. If this reduction were produced by mechanical means the heat emitted could not change the temperature of the body as many degrees as the cooling. It would, however, require the use of a force which would certainly be very considerable. Since solid bodies are susceptible to but small changes of temperature by changes of volume, and since, moreover, the condition for the best use of heat in the development of motive power is that any change of temperature should be due to a change of volume, solid bodies do not seem to be well adapted to realize this power.

This is equally true in the case of liquids ; the same reasons could be given for rejecting them.*

We shall not speak here of the practical difficulties, which are innumerable. The movements produced by the expansion and compression of solids or liquids can only be very small. To extend these movements we should be forced to use complicated

* The recent experiments of M. Oersted on the compressibility of water have shown that for a pressure of 5 atmospheres the temperature of the liquid undergoes no perceptible change. (See *Annales de Chimie et de Physique*, February, 1823, p. 192.)

mechanisms and also materials of the greatest strength to transmit enormous pressures; and, finally, the successive operations
could only proceed very slowly compared with those of the
ordinary heat-engine, so that even large and expensive machines would produce only insignificant results.

Elastic fluids, gases, or vapors are the instruments peculiarly fitted for the development of the motive power of heat; they
unite all the conditions necessary for this service; they may be
easily compressed, and possess the property of almost indefinite
expansion; changes of volume occasion in them great changes
of temperature, and finally they are very mobile, can be easily
and quickly heated and cooled, and readily transported from
one place to another, so that they are able to produce rapidly
the effects expected of them.

We can easily conceive of many machines fitted for the development of the motive power of heat by the use of elastic
fluids, but however they are constructed in other respects, the
following conditions must not be lost sight of:

1. The temperature of the fluid should first be raised to the
highest degree possible, in order to obtain a great descent of
caloric and consequently a great production of motive power.

2. For the same reason the temperature of the refrigerator
should be as low as possible.

3. The operations must be so conducted that the transfer of
the elastic fluid from the highest to the lowest temperature
should be due to an increase of volume—that is, that the cooling of the gas should occur spontaneously by the effect of expansion.

The limits to which the temperature of the fluid can be
raised in the first operation are determined only by the temperature of combustion; they are very much higher than ordinary temperatures. The limits of cooling are reached in the
temperature of the coldest bodies which we can conveniently
use in large quantities; the body most used for this purpose is
the water available at the place where the operation is carried on.

As to the third condition, it introduces difficulties in the
realization of the motive power of heat, when the object is to
profit by great differences of temperature, that is to utilize
great descents of caloric. For in that case the gas must change
from a very high temperature to a very low one, by expansion,

which requires a great change of volume and density. To effect this the gas must at first be subjected to a very high pressure, or it must acquire by expansion an enormous volume, either of which conditions is difficult to realize. The first necessitates the use of very strong vessels to contain the gas when it is at a high pressure and temperature; the second requires the use of vessels of a very large size.

In fact, these are the principal obstacles in the way of profitably using in steam-engines a large portion of the motive power of heat. We are of necessity limited to the use of a small descent of caloric, although the combustion of coal furnishes us with the means of obtaining a very great one. In the use of steam-engines the elastic fluid is rarely developed at a pressure higher than 6 atmospheres, which pressure corresponds to nearly 160 degrees centigrade, and condensation is rarely effected at a temperature much below 40 degrees; the descent of caloric from 160 to 40 degrees is 120 degrees, while we can obtain by combustion a descent of from 1000 to 2000 degrees.

To conceive of this better, we shall recall what we have previously called the descent of caloric : It is the transfer of heat from a body, A, at a high temperature to a body, B, whose temperature is lower. We say that the descent of caloric is 100 degrees or 1000 degrees when the difference of temperature between the bodies A and B is 100 or 1000 degrees. In a steam-engine working under a pressure of 6 atmospheres the temperature of the boiler is 160 degrees. This is the temperature of the body A ; it is maintained by contact with the furnace at a constant temperature of 160 degrees, and affords a continual supply of the heat necessary to the formation of steam.

The condenser is the body B; it is maintained by means of a current of cold water at an almost constant temperature of 40 degrees, and continually absorbs the caloric which is carried to it by the steam from the body A. The difference of temperature between these two bodies is 160—40, or 120 degrees ; it is for this reason that we say that the descent of caloric is in this case 120 degrees.

Coal is capable of producing by combustion a higher temperature than 1000 degrees, and the temperature of the cold water which we ordinarily use is about 10 degrees, so that we can easily obtain a descent of caloric of 1000 degrees, of which only

120 degrees are utilized by steam-engines, and even these 120 degrees are not all used to advantage ; there are always considerable losses due to useless re-establishments of equilibrium in the caloric.

It is now easy to perceive the advantage of those engines which are called high-pressure engines over those in which the pressure is lower : *this advantage depends essentially upon the power of utilizing a larger descent of caloric.* The steam being produced under greater pressure is also at a higher temperature, and as the temperature of condensation is always nearly the same the descent of caloric is evidently greater.

But to obtain the most favorable results from high-pressure engines the descent of caloric must be used to the greatest advantage. It is not enough that the steam should be produced at a high temperature, but it is also necessary that it should attain a sufficiently low temperature by its expansion alone. It should thus be the characteristic of a good steam-engine not only that it uses the steam under high pressure, *but that it uses it under successive pressures which are very variable, very different from each other, and progressively decreasing.**

* This principle, which is the real basis of the theory of the steam-engine, has been developed with great clearness by M. Clément in a memoir presented to the Academy of Sciences a few years ago. This memoir has never been printed, and I owe my acquaintance with it to the kindness of the author. In it not only is this principle established, but applied to various systems of engines actually in use ; the motive power of each is calculated by the help of the law cited on p. 35 and compared with the results of experiment. This principle is so little known or appreciated that Mr. Perkins, the well-known London mechanician, recently constructed an engine in which the steam, formed under a pressure of 35 atmospheres, a pressure never before utilized, experienced almost no expansion, as we may easily be convinced by the slightest knowledge of the engine. It is composed of a single cylinder, which is very small, and at each stroke is entirely filled with steam formed under a pressure of 35 atmospheres. The steam does no work by expansion, for there is no room for the expansion to take place ; it is condensed as soon as it passes out of the small cylinder. It acts only under a pressure of 35 atmospheres, and not, as the best usage would require, under progressively decreasing pressures. This engine of Mr. Perkins does not realize the hopes which it at first excited. It was claimed that the economy of coal in this machine was $\frac{9}{10}$ greater than in the best machines of Watt, and that it also possessed other advantages over them. (See *Annales de Chimie et de Physique*, April, 1823, p. 429.) These assertions have not been verified. Mr. Perkins's engine may nevertheless be considered a valuable invention in that it has proved it to be

In order to show, to a certain extent, *a posteriori* the advantage of high-pressure engines, let us assume that the steam formed under atmospheric pressure is contained in a cylindri-

feasible to use steam under much higher pressures than ever before, and because when properly modified it may lead us to really useful results.

Watt, to whom we owe almost all the great improvements in the steam-engine, and who has brought these machines to a state of perfection which can hardly be surpassed, was the first to use steam under progressively decreasing pressures. In many cases he checked the introduction of the steam into the cylinder at one-half, one-third, or one-quarter of the stroke of the piston, which was thus completed under a pressure which constantly diminished. The first engines working on this principle date from 1778. Watt had conceived the idea in 1769, and took out a patent for it in 1782.

A table annexed to Watt's patent is here presented. In it he supposes the vapor to enter the cylinder during the first quarter of the stroke of the piston, and he then calculates the mean pressure by dividing the stroke into twenty parts:

PARTS OF THE PATH FROM THE HEAD OF THE CYLINDER		DECREASING PRESSURE OF THE VAPOR, THE TOTAL PRESSURE BEING 1	
	0.05	1.000	
	0.10	1.000	
	0.15	Steam entering freely from the boiler.	1.000	Total pressure.
	0.20	1.000	
Quarter......0.25		1.000	
	0.30	0.830	
	0.35	0.714	
	0.40	0.625	
	0.45	0.555	
Half.........0.50		0.500..Half the original pressure.	
	0.55	0.454	
	0.60	The steam cut off, and moving the piston by expansion alone.	0.417	
	0.65	0.385	
	0.70	0.375	
	0.75	0.333..One-third.	
	0.80	0.312	
	0.85	0.294	
	0.90	0.277	
Bottom of	0.95	0.262	
cylinder..1.00		0.250..One-quarter.	

Total.........11.583

Mean pressure, $\dfrac{11.583}{20} = 0.579.$

On this showing he remarks that the mean pressure is more than half of the original pressure, so that a quantity of steam equal to one-quarter will produce an effect greater than one-half [*freely introduced from the boiler until the end of the stroke*].

Watt here assumes that the expansion of the steam is in accordance with Mariotte's law. This assumption should not be considered correct, be-

52

cal vessel, *abcd* (Fig. 5). under the piston *cd*, which at first
touches the base *ab;* the steam, after moving the piston from
ab to *cd*, will subsequently act in a manner with
which we need not occupy ourselves. Let us
suppose that after the piston has reached *cd* it is
forced down to *ef* without escape of steam, or
loss of any of its caloric. It will be compressed
into the space *abef*, and its density, elastic force,
and temperature will all increase together.

If the steam, instead of being formed under
atmospheric pressure, were produced exactly in
the state in which it is when compressed into
abef, and if, after having moved the piston from
ab to *ef* by its introduction into the cylinder, it

Fig. 5

should move it from *ef* to *cd* solely by expansion, the motive
power produced would be greater than in the first case. In
fact, an equal movement of the piston would take place under
the influence of a higher pressure, although this would be variable and even progressively decreasing.

The steam would require for its formation a precisely equal
quantity of caloric, but this caloric would be used at a higher
temperature.

It is from considerations of this kind that engines with two
cylinders (compound engines) were introduced, which were in-
vented by Mr. Hornblower and improved by Mr. Woolf. With

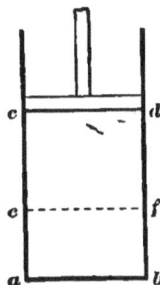

cause, on the one hand the temperature of the elastic fluid is lowered by expansion, and on the other there is nothing to show that a part of this fluid
does not condense by expansion. Watt should also have taken into account the force necessary to expel the steam remaining after condensation, whose quantity is greater in proportion as the expansion has been
carried further. Dr. Robinson added to Watt's work a simple formula
to calculate the effect of the expansion of steam, but this formula is affected by the same errors to which we have just called attention. It has,
however, been useful to constructors in furnishing them with a means of
calculation sufficiently exact to be of use in practice. We have thought it
worth while to recall these facts because they are little known, especially
in France. Engines have been constructed there after the models of inventors but without much appreciation of the principles on which these
models were made. The neglect of these principles has often led to grave
faults. Engines which were originally well conceived have deteriorated
in the hands of unskilful constructors, who, wishing to introduce unimportant improvements, have neglected fundamental considerations which
they did not know enough to appreciate.

respect to the economy of fuel, they are considered the best engines. They are composed of a small cylinder, which at each stroke of the piston is more or less and often entirely filled with steam, and of a second cylinder, of a capacity usually four times as great, which receives only the steam which has already been used in the first one. Thus the volume of the steam at the end of this operation is at least four times its original volume. It is carried from the second cylinder directly into the condenser; but it is evident that it could be carried into a third cylinder four times as large as the second, where its volume would become sixteen times its original volume. The chief obstacle to the use of a third cylinder of this kind is the large space which it requires, and the size of the openings which are necessary to allow the steam to escape.*

We shall say nothing more on this subject, our object not being to discuss the details of construction of heat-engines. These should be treated in a separate work. No such work exists at present, at least in France.†

* It is easy to perceive the advantages of having two cylinders, for when there is only one the pressure on the piston will vary very much between the beginning and end of the stroke. Also, all the portions of the machine designed to transmit the action must be strong enough to resist the first impulse, and fitted together perfectly so as to avoid sudden motions by which they might be damaged and which would soon wear them out. This would be specially true of the walking-beam, the supports, the connecting-rod, the crank, and of the first cog-wheels. In these parts the irregularity of the impulse would be most felt and would do the most damage. The steam-chest would also have to be strong enough to resist the highest pressure employed, and large enough to contain the vapor when its volume is increased. If two cylinders are used the capacity of the first need not be great, so that it is easy to give it the strength required, while the second must be large but need not be very strong.

Engines with two cylinders have been planned on proper principles but have often fallen far short of yielding as good results as might have been expected of them. This is the case principally because the dimensions of the different parts are difficult to arrange and are often not in proper proportion to each other. There are no good models of these engines, while there are excellent ones of those constructed after Watt's plan. To this is due the irregularity which we observe in the effects produced by the former, while those produced by the latter are almost uniform.

† In the work entitled *De la Richesse Minéral*, by M. Héron de Villefosse, vol. iii., p. 50 *seq.*, we find a good description of the steam-engines now used in mining. The subject has been treated with sufficient fulness in England in the *Encyclopædia Britannica*. Some of the data which we have employed have been taken from the latter work.

While the expansion of the steam is limited by the dimensions of the vessels in which it dilates, the degree of condensation at which it is possible to begin to use it is only limited by the resistance of the vessels in which it is generated—namely, of the boilers. In this respect we are far from having reached the possible limits. The character of the boilers in general use is altogether bad; although the tension of the steam is rarely carried in them beyond 4 to 6 atmospheres, they often burst and have caused serious accidents. It is no doubt quite possible to avoid such accidents and at the same time to make the tension of the steam greater than that commonly employed.

Besides the high-pressure engines with two cylinders of which we have been speaking, there are also high-pressure engines with one cylinder. Most of these have been constructed by two skilful English engineers, Messrs. Trevithick and Vivian. They use the steam under a very high pressure, sometimes 8 or 10 atmospheres, but they have no condenser. The steam, after its entrance into the cylinder, undergoes a certain expansion, but its pressure is always greater than that of the atmosphere. When it has done its work, it is ejected into the atmosphere. It is evident that this mode of procedure is entirely equivalent, with respect to the motive power produced, to condensing the steam at 100 degrees, and that we lose a part of the useful effect, but engines thus worked can dispense with the condenser and air-pump. They are less expensive than the others, and are not so complicated; they take less room, and can be used where it is not possible to obtain a current of cold water sufficient to effect condensation. In such places they possess an incalculable advantage, since no others can be used. They are used principally in England to draw wagons for the carriage of coal on railroads, either in the interior of mines or on the surface.

Some remarks may still be made on the use of permanent gases and vapors other than water vapor in the development of the motive power of heat.

Various attempts have been made to produce motive power by the action of heat on atmospheric air. This gas, in comparison with water vapor, presents some advantages and some disadvantages, which we shall now examine.

1. It has this notable advantage over water vapor, that since for the same volume it has a much smaller capacity for

heat it cools more for an equal expansion, as is proved by what we have previously said. We have seen the importance of effecting the greatest possible changes of temperature by changes of volume alone.

2. Water vapor can be formed only by the aid of a boiler, while atmospheric air can be heated directly by combustion within itself. Thus a considerable loss is avoided, not only in the quantity of heat, but also in its thermometric degree. This advantage belongs exclusively to atmospheric air; the other gases do not possess it; they would be even more difficult to heat than water vapor.

3. In order to produce a great expansion of the air, and to cause thereby a great change of temperature, it would be necessary to subject it in the first place to rather a high pressure, to compress it by an air-pump or by some other means before heating it. This operation would require a special apparatus which does not form a part of the steam-engine. In it the water is in a liquid state when it enters the boiler, and requires only a small force-pump to introduce it.

4. The cooling of the vapor by the contact of the refrigerating body is more rapid and easy than the cooling of air could be. It is true that we have the resource of ejecting it into the atmosphere. This procedure would have the further advantage that we could then dispense with a refrigerator, which is not always at our disposal, but in that case the air must not expand so far that its pressure is lower than that of the atmosphere.

5. One of the most serious drawbacks to the employment of steam is that it cannot be used at high temperatures except with vessels of extraordinary strength. This is not true of air, for which there is no necessary relation between its temperature and elastic force. The air, then, seems better fitted than steam to realize the motive power of the descent of caloric at high temperatures; perhaps at low temperatures water vapor would be preferable. We can even conceive of the possibility of making the same heat act successively in air and in water vapor. All that would be necessary would be to keep the temperature of the air sufficiently high, after it had been used, and instead of ejecting it immediately into the atmosphere, to surround a steam-boiler with it, as if it came directly from the fire-box.

The use of atmospheric air for the development of the motive power of heat presents very great practical difficulties which, however, may not be insurmountable. These difficulties once overcome, it will doubtless be far superior to water vapor.*

As for other permanent gases, they should be finally rejected; they have all the inconveniences of atmospheric air without any of its advantages.

The same may be said of other vapors in comparison with water vapor.

*Among the attempts made to develop the motive power of heat by the use of atmospheric air, we should notice particularly those of MM. Niepce, which were made in France several years ago by means of an apparatus, called by the inventors *pyréolophore*. This instrument consists essentially of a cylinder, furnished with a piston, and filled with atmospheric air at ordinary density. Into this is projected some combustible substance in a highly attenuated form, which remains in suspension for a moment in the air and is then ignited. The combustion produces nearly the same effect as if the elastic fluid were a mixture of air and combustible gas—of air and carburetted hydrogen, for example—a sort of explosion occurs and a sudden expansion of the elastic fluid, which is made use of by causing it to act altogether against the piston. This moves through a certain distance, and the motive power is thus realized. There is nothing to prevent a renewal of the air and a repetition of the first operation. This very ingenious engine, which is especially interesting on account of the novelty of its principle, fails in an essential particular. The substance used for the combustible (lycopodium powder, that which is used to produce flames on the stage) is so expensive, that all other advantages are outweighed, and unfortunately it is difficult to make use of a moderately cheap combustible, for it requires a substance that is very finely pulverized, in which the ignition is prompt, is propagated rapidly, and which leaves little or no residue.

Instead of following MM. Niepce's operations it would seem to us better to compress the air by air-pumps and to conduct it through a perfectly sealed fire-box into which the combustible is introduced in small quantities by some mechanism which is easy to conceive of; to allow it to develop its action in a cylinder with a piston or in any other envelope capable of enlargement; to eject it finally into the atmosphere, or even to pass it under a steam-boiler in order to utilize its remaining heat.

The chief difficulties which we should have to meet in this mode of operation would be the enclosure of the fire-box in a sufficiently solid envelope, the suitable control of the combustion, the maintenance of a moderate temperature in the several parts of the engine, and the prevention of a rapid deterioration of the cylinder and piston. We do not consider these difficulties insurmountable.

It is said that successful attempts have been made in England to develop motive power by the action of heat on atmospheric air. We do not know what these are, if, indeed, they have really been made.

It would no doubt be preferable if there were an abundant supply of a liquid which evaporated at a higher temperature than water, the specific heat of whose vapor was less for equal volume, and which did not injure the metals used in the construction of an engine ; but no such body exists in nature.

The use of alcohol vapor has been suggested, and engines have even been constructed in order to make it possible, in which the mixture of the vapor with the water of condensation is avoided by applying the cold body externally instead of introducing it into the engine.

It was thought that alcohol vapor possessed a notable advantage on account of its having a greater tension than that of water vapor at the same temperature. We see in this only another difficulty to be overcome. The principal defect of water vapor is its excessive tension at a high temperature, and this defect is still more marked in alcohol vapor. As for the advantage which it was believed to have with respect to a larger production of motive power, we know from the principles stated above that they are imaginary.

Thus it is with the use of water vapor and atmospheric air that the future attempts to improve the steam-engine should be made. All efforts should be directed to utilize by means of these agents the largest possible descents of caloric.

We shall conclude by showing how far we are from the realization, by means already known, of all the motive power of the combustibles.

A kilogram of coal burned in the calorimeter furnishes a quantity of heat capable of raising the temperature of about 7000 kilograms of water 1 degree—that is, from the definition given (page 43) it furnishes 7000 units of heat. The largest descent of caloric which can be realized is measured by the difference of the temperature produced by combustion and that of the refrigerating body. It is difficult to see any limit to the temperature of combustion other than that at which the combination of the combustible with oxygen is effected. Let us assume, however, that this limit is 1000 degrees, which is certainly within the bounds of truth. We shall assume the temperature of the refrigerator to be 0 degrees.

We have calculated approximately (page 45) the quantity of motive power developed by 1000 units of heat in passing from the temperature 100 to the temperature 99, and have found it to be 1.112 units, each equal to 1 meter of water raised 1 meter.

If the motive power were proportional to the descent of caloric, if it were the same for each thermometric degree, nothing would be easier than to estimate it from 1000 to 0 degrees. Its value would be

$$1.112 \times 1000 = 1112.$$

But as this law is only approximate, and perhaps at high temperatures departs a good deal from the truth, we can only make a very rough estimate. Let us suppose the number 1112 to be reduced one-half—that is, to 560.

Since one kilogram of coal produces 7000 units of heat, and since the number 560 is referred to 1000 units, we must multiply it by 7, which gives us

$$7 \times 560 = 3920,$$

which is the motive power of one kilogram of coal.

In order to compare this theoretical result with the results of experiment, we shall inquire how much motive power is actually developed by one kilogram of coal in the best heat-engines known.

The engines which have thus far offered the most advantageous results are the large engines with two cylinders used in the pumping out of the tin and copper mines of Cornwall. The best results which they have furnished are as follows: Sixty-five million pounds of water have been raised one English foot by the burning of one bushel of coal (the weight of a bushel is 88 lbs.). This result is equivalent to raising 195 cubic meters of water one meter by the use of one kilogram of coal, which consequently produces 195 units of motive power.* ·

* The result given here was furnished by an engine whose large cylinder was 35 inches in diameter, with a 7-foot stroke ; it is used to pump out one of the mines of Cornwall, called "Wheal Abraham." This result should in a way be considered as an exception, for it only was accomplished for a short time during one month. A product of 30 million lbs. raised one English foot by a bushel of coal weighing 88 lbs. is generally considered to be an excellent result for a steam-engine. It is sometimes reached by the engines made on Watt's system, but has rarely been exceeded. This result expressed in French units is equal to 104000 kilograms raised one meter by the burning of one kilogram of coal.

By what we ordinarily understand as one horse-power in the calculation

195 units are only one-twentieth of 3920, the theoretical maximum; consequently only $\frac{1}{20}$ of the motive power of the combustible has been utilized.

We have, moreover, chosen our example from among the best steam-engines known. Most of the others have been very inferior. For example, Chaillot's engine raises 20 cubic meters of water 33 meters in consuming 30 kilograms of coal, which is equivalent to 22 units of motive power to 1 kilogram, a result nine times less than that cited above, and one hundred and eighty times less than the theoretical maximum.

We should not expect ever to employ in practice all the motive power of the combustibles used. The efforts which one would make to attain this result would be even more harmful than useful if they led to the neglect of other important considerations. The economy of fuel is only one of the conditions which should be fulfilled by steam-engines; in many cases it is only a secondary consideration. It must often yield the precedence to safety, to the solidity and durability of the engine, to the space which it must occupy, to the cost of its construction, etc. To be able to appreciate justly in each case the considerations of convenience and economy which present themselves, to be able to recognize the most important of those which are only subordinate, to adjust them all suitably, and finally to reach the best result by the easiest method—such should be the power of the man who is called on to direct and co-ordinate the labors of his fellow-men, and to make them concur in attaining a useful purpose.

Biographical Sketch

Nicolas-Léonard-Sadi Carnot was born in Paris on June 1, 1796 ; the son of the illustrious engineer, soldier, and statesman who played so prominent a part in the history of France during the Revolution. He was educated at the École Poly-

---

of the efficiency of steam-engines, a 10 horse-power engine should raise 10×75, or 750 kilograms 1 meter in a second, or $750 \times 3600 = 2700000$ kilograms 1 meter in an hour.

If we suppose each kilogram of coal to raise 104000 kilograms, it is necessary to divide 2700000 by 104000 to find the coal burned in one hour by the 10 horse-power engine, which gives us $\frac{2700}{104} = 26$ kilograms. But it is very rare that a 10 horse-power engine consumes less than 26 kilograms of coal an hour.

technique, and served for several years as an officer of engineers and on the general staff. His inclinations towards the study of science were so strong that they controlled the whole course of his life. While still engaged in his profession he devoted such time as he could spare to scientific investigations, and he at last resigned from the army in order to obtain more leisure for studious pursuits. He died of the cholera on August 24, 1832. The memoir on the "Motive Power of Heat" is the only one which he published. It shows that he possessed a mind able to penetrate to the heart of a question, and to invent general methods of reasoning. The extracts from his note-book, published by his brother, indicate that he was also fertile in devising experiments. It is interesting to note that the doubt of the validity of the substantial theory of heat, expressed by him in his memoir, developed later into complete disbelief, and that he not only adopted the mechanical theory of heat, but planned experiments to test it similar to those of Joule, and calculated that the mechanical equivalent of heat is equal to 370 kilogram-meters.

ON THE MOTIVE POWER OF HEAT, AND ON THE LAWS WHICH CAN BE DEDUCED FROM IT FOR THE THEORY OF HEAT

BY

R. CLAUSIUS

(Poggendorff's *Annalen*, vol. lxxix., pp. 376 and 500.   1850)

CONTENTS

ON THE MOTIVE POWER OF HEAT, AND ON THE LAWS WHICH CAN BE DEDUCED FROM IT FOR THE THEORY OF HEAT

BY

R. CLAUSIUS

SINCE heat was first used as a motive power in the steam-engine, thereby suggesting from practice that a certain quantity of work may be treated as equivalent to the heat needed to produce it, it was natural to assume also in theory a definite relation between a quantity of heat and the work which in any possible way can be produced by it, and to use this relation in drawing conclusions about the nature and the laws of heat itself. In fact, several fruitful investigations of this sort have already been made; yet I think that the subject is not yet exhausted, but on the other hand deserves the earnest attention of physicists, partly because serious objections can be raised to the conclusions that have already been reached, partly because other conclusions, which may readily be drawn and which will essentially contribute to the establishment and completion of the theory of heat, still remain entirely unnoticed or have not yet been stated with sufficient definiteness.

The most important of the researches here referred to was that of S. Carnot,* and the ideas of this author were afterwards given analytical form in a very skilful way by Clapeyron.† Carnot showed that whenever work is done by heat and

---

* *Réflexions sur la puissance motrice du feu, et sur les machines propres à développer cette puissance, par S. Carnot. Paris*, 1824. I have not been able to obtain a copy of this book, and am acquainted with it only through the work of Clapeyron and Thomson, from the latter of whom are quoted the extracts afterwards given.

† *Journ. de l'École Polytechnique*, vol. xix. (1834), and Pogg. *Ann.*, vol. lix.

no permanent change occurs in the condition of the working
body, a certain quantity of heat passes from a hotter to a colder
body. In the steam - engine, for example, by means of the
steam which is developed in the boiler and precipitated in the
condenser, heat is transferred from the grate to the condenser.
This *transfer* he considered as the heat change, corresponding
to the work done. He says expressly that no heat is lost in the
process, but that the *quantity of heat* remains unchanged, and
adds : " This fact is not doubted ; it was assumed at first without investigation, and then established in many cases by calori-
metric measurements. To deny it would overthrow the whole
theory of heat, of which it is the foundation." I am not aware,
however, that it has been sufficiently proved by experiment
that no loss of heat occurs when work is done ; it may, perhaps,
on the contrary, be asserted with more correctness that even if
such a loss has not been proved directly, it has yet been shown
by other facts to be not only admissible, but even highly probable. If it be assumed that heat, like a substance, cannot
diminish in quantity, it must also be assumed that it cannot
increase. It is, however, almost impossible to explain the heat
produced by friction except as an increase in the quantity of
heat. The careful investigations of Joule, in which heat is
produced in several different ways by the application of mechanical work, have almost certainly proved not only the possibility of increasing the quantity of heat in any circumstances
but also the law that the quantity of heat developed is proportional to the work expended in the operation. To this it must
be added that other facts have lately become known which
support the view, that heat is not a substance, but consists in a
motion of the least parts of bodies. If this view is correct, it
is admissible to apply to heat the general mechanical principle
that a motion may be transformed into work, and in such a
manner that the loss of *vis viva* is proportional to the work accomplished.

These facts, with which Carnot also was well acquainted, and
the importance of which he has expressly recognized, almost
compel us to accept the equivalence between heat and work, on
the modified hypothesis that the accomplishment of work requires not merely a change in the distribution of heat, but also
an actual consumption of heat, and that, conversely, heat can
be developed again by the expenditure of work.

THE SECOND LAW OF THERMODYNAMICS

In a memoir recently published by Holtzmann,* it seems at first as if the author intended to consider the matter from this latter point of view. He says (p. 7) : " The action of the heat supplied to the gas is either an elevation of temperature, in conjunction with an increase in its elasticity, or mechanical work, or a combination of both, and the mechanical work is the equivalent of the elevation of temperature. The heat can only be measured by its effects ; of the two effects mentioned the mechanical work is the best adapted for this purpose, and it will accordingly be so used in what follows. I call the unit of heat the heat which by its entrance into a gas can do the mechanical work a—that is, to use definite units, which can lift a kilograms through 1 meter." Later (p. 12) he also calculates the numerical value of the constant a in the same way as Mayer had already done,† and obtains a number which corresponds with the heat equivalent obtained by Joule in other entirely different ways. In the further extension of his theory, however, in particular in the development of the equations from which his conclusions are drawn, he proceeds exactly as Clapeyron did, so that in this part of his work he tacitly assumes that the quantity of heat is constant.

The difference between the two methods of treatment has been much more clearly grasped by W. Thomson, who has extended Carnot's discussion by the use of the recent observations of Regnault on the tension and latent heat of water vapor. ‡ He speaks of the obstacles which lie in the way of the unrestricted assumption of Carnot's theory, calling special attention to the researches of Joule, and also raises a fundamental objection which may be made against it. Though it may be true in any case of the production of work, when the working body has returned to the same condition as at first, that heat passes from a warmer to a colder body, yet on the other hand it is not generally true that whenever heat is transferred work is done. Heat can be transferred by simple conduction, and in all such cases, if the mere transfer of heat were the true equivalent of work, there would be a loss of working power in Nature, which is hardly conceivable. Nevertheless, he concludes that in the

* *Ueber die Wärme und Elasticität der Gase und Dämpfe*, von C. Holtzmann, Mannheim, 1845 ; also Pogg. *Ann.*, vol. 72a.

† *Ann. der Chem. und Pharm.* of Wöhler and Liebig, vol. xlii., p. 239.

‡ *Transactions of the Royal Society of Edinburgh*, vol. xvi.

present state of the science the principle adopted by Carnot is still to be taken as the most probable basis for an investigation of the motive power of heat, saying : " If we abandon this principle, we meet with innumerable other difficulties—insuperable without further experimental investigation, and an entire reconstruction of the theory of heat from its foundation."[*]

I believe that we should not be daunted by these difficulties, but rather should familiarize ourselves as much as possible with the consequences of the idea that heat is a motion, since it is only in this way that we can obtain the means wherewith to confirm or to disprove it. Then, too, I do not think the difficulties are so serious as Thomson does, since even though we must make some changes in the usual form of presentation, yet I can find no contradiction with any proved facts. It is not at all necessary to discard Carnot's theory entirely, a step which we certainly would find it hard to take, since it has to some extent been conspicuously verified by experience. A careful examination shows that the new method does not stand in contradiction to the essential principle of Carnot, but only to the subsidiary statement *that no heat is lost*, since in the production of work it may very well be the case that at the same time a certain quantity of heat is consumed and another quantity transferred from a hotter to a colder body, and both quantities of heat stand in a definite relation to the work that is done. This will appear more plainly in the sequel, and it will there be shown that the consequences drawn from the two assumptions are not only consistent with one another, but are even mutually confirmatory.

1. CONSEQUENCES OF THE PRINCIPLE OF THE EQUIVALENCE OF HEAT AND WORK

We shall not consider here the kind of motion which can be conceived of as taking place within bodies, further than to assume in general that the particles of bodies are in motion, and that their heat is the measure of their *vis viva*, or rather still more generally, we shall only lay down a principle conditioned by that assumption as a fundamental principle, in the words : In all cases in which work is produced by the agency of heat, a quantity of heat is consumed which is proportional to the

[*] *Math. and Phys. Papers*, vol. i., p. 119, note.

work done ; and, conversely, by the expenditure of an equal quantity of work an equal quantity of heat is produced.

Before we proceed to the mathematical treatment of this principle, some immediate consequences may be premised which affect our whole method of treatment, and which may be understood without the more definite demonstration which will be given them later by our calculations.

It is common to speak of the *total heat* of bodies, especially of gases and vapors, by which term is understood the sum of the free and latent heat, and to assume that this is a quantity dependent only on the actual condition of the body considered, so that, if all its other physical properties, its temperature, its density, etc., are known, the total heat contained in it is completely determined. This assumption, however, is no longer admissible if our principle is adopted. Suppose that we are given a body in a definite state—for example, a quantity of gas with the temperature t_0 and the volume v_0—and that we subject it to various changes of temperature and volume, which are such, however, as to bring it at last to its original state again. According to the common assumption, its total heat will again be the same as at first, from which it follows that if during one part of its changes heat is communicated to it from without, the same quantity of heat must be given up by it in the other part of its changes. Now with every change of volume a certain amount of work must be done by the gas or upon it, since by its expansion it overcomes an external pressure, and since its compression can be brought about only by an exertion of external pressure. If, therefore, among the changes to which it has been subjected there are changes of volume, work must be done upon it and by it. It is not necessary, however, that at the end of the operation, when it is again brought to its original state, the work done by it shall on the whole equal that done upon it, so that the two quantities of work shall counterbalance each other. There may be an excess of one or the other of these quantities of work, since the compression may take place at a higher or lower temperature than the expansion, as will be more definitely shown later on. To this excess of work done by the gas or upon it there must correspond, by our principle, a proportional excess of heat consumed or produced, and the gas cannot give up to the surrounding medium the same amount of heat as it receives.

The same contradiction to the ordinary assumption about the *total heat* may be presented in another way. If the gas at t_0 and v_0 is brought to the higher temperature t_1 and the larger volume v_1, the quantity of heat which must be imparted to it is, on that assumption, independent of the way in which the change is brought about; from our principle, however, it is different, according as the gas is first heated while its volume, v_0, is constant, and then allowed to expand at the constant temperature t_1, or is first expanded at the constant temperature t_0, and then heated, or as the expansion and heating are interchanged in any other way or even occur together, since in all these cases the work done by the gas is different.

In the same way, if a quantity of water at the temperature t_0 is changed into vapor at the temperature t_1 and of the volume v_1, it will make a difference in the amount of heat needed if the water as such is first heated to t_1 and then evaporated, or if it is evaporated at t_0 and the vapor then brought to the required volume and temperature. v_1 and t_1, or finally if the evaporation occurs at any intermediate temperature.

From these considerations and from the immediate application of the principle, it may easily be seen what conception must be formed of *latent* heat. Using again the example already employed, we distinguish in the quantity of heat which must be imparted to the water during its changes the *free* and the *latent* heat. Of these, however, we may consider only the former as really present in the vapor that has been formed. The latter is not merely, as its name implies, *concealed* from our perception, but it is *nowhere present ;* it is *consumed* during the changes in doing work.

In the heat consumed we must still introduce a distinction—that is to say, the work done is of two kinds. First, there is a certain amount of work done in overcoming the mutual attractions of the particles of the water, and in separating them to such a distance from one another that they are in the state of vapor. Secondly, the vapor during its evolution must push back an external pressure in order to make room for itself. The former work we shall call the *internal*, the latter the *external* work, and shall partition the latent heat accordingly.

It can make no difference with respect to the *internal* work whether the evaporation goes on at t_0 or at t_1, or at any intermediate temperature, since we must consider the attractive force of the particles, which is to be overcome, as invariable.* The *external* work, on the other hand, is regulated by the pressure as dependent on the temperature. Of course the same is true in general as in this special example, and therefore if it was said above that the quantity of heat which must be imparted to a body, to bring it from one condition to another, depended not merely on its initial and final conditions, but also on the way in which the change takes place, this statement refers only to that part of the *latent* heat which corresponds to the *external* work. The other part of the *latent* heat, as also the *free* heat, are independent of the way in which the changes take place.

If now the vapor at t_1 and v_1 is again transformed into water, work will thereby be *expended*, since the particles again yield to their attractions and approach each other, and the external pressure again advances. Corresponding to this, heat must be *produced*, and the so-called liberated heat which appears during the operation does not merely come out of concealment but is actually made new. The heat produced in this reversed operation need not be equal to that used in the direct one, but that part which corresponds to the *external* work may be greater or less according to circumstances.

We shall now turn to the mathematical discussion of the subject, in which we shall restrict ourselves to the consideration of the *permanent gases* and of *vapors at their maximum density*, since these cases, in consequence of the extensive knowledge we have of them, are most easily submitted to calculation, and besides that are the most interesting.

Let there be given a certain quantity, say a unit of weight, of a *permanent gas*. To determine its present condition, three magnitudes must be known: the pressure upon it, its volume,

---

* It cannot be raised, as an objection to this statement, that water at t_1 has less cohesion than at t_0, and that therefore less work would be needed to overcome it. For a certain amount of work is used in diminishing the cohesion, which is done while the water as such is heated, and this must be reckoned in with that done during the evaporation. It follows at once that only a part of the heat, which the water takes up from without while it is being heated, is to be considered as free heat, while the remainder is used in diminishing the cohesion. This view is also consistent with the circumstance that water has so much greater a specific heat than ice, and probably also than its vapor.

and its temperature. These magnitudes are in a mutual relationship, which is expressed by the combined laws of Mariotte and Gay-Lussac*, and may be represented by the equation:

(I.) $$pv = R\,(a+t),$$

where p, v, and t represent the pressure, volume, and temperature of the gas in its present condition, a is a constant, the same for all gases, and R is also a constant, which in its complete form is $\frac{p_0\,v_0}{a+t_0}$, if p_0, v_0, and t_0 are the corresponding values of the three magnitudes already mentioned for any other condition of the gas. This last constant is in so far different for the different gases that it is inversely proportional to their specific gravities.

It is true that Regnault has lately shown, by a very careful investigation, that this law is not strictly accurate, yet the departures from it are in the case of the permanent gases very small, and only become of consequence in the case of those gases which can be condensed into liquids. From this it seems to follow that the law holds with greater accuracy the more removed the gas is from its condensation point with respect to pressure and temperature. We may therefore, while the accuracy of the law for the permanent gases in their ordinary condition is so great that it can be treated as complete in most investigations, think of a limiting condition for each gas, in which the accuracy of the law is actually complete. We shall, in what follows, when we treat the permanent gases as such, assume this ideal condition.

According to the concordant investigations of Regnault and Magnus, the value of $\frac{1}{a}$ for atmospheric air is equal to 0.003665, if the temperature is reckoned in centigrade degrees from the freezing-point. Since, however, as has been mentioned, the gases do not follow the M. and G. law exactly, the same value of $\frac{1}{a}$ will not always be obtained, if the measurements are made in different circumstances. The number here given holds for the case when air is taken at 0° under the pressure of *one* atmosphere, and heated to 100° at constant volume, and the

---

* This law will hereafter, for brevity, be called the M. and G. law, and Mariotte's law will be called the M. law.

increase of its expansive force observed. If, on the other hand, the pressure is kept constant, and the increase of its volume observed, the somewhat greater number 0.003670 is obtained. Further, the numbers increase if the experiment is tried under a pressure higher than the atmospheric pressure, while they diminish somewhat for lower pressures. It is not therefore possible to decide with certainty on the number which should be adopted for the gas in the ideal condition in which naturally all differences must disappear; yet the number 0.003665 will surely not be far from the truth, especially since this number very nearly obtains in the case of hydrogen, which probably approaches the most nearly of all the gases the ideal condition, and for which the changes are in the opposite sense to those of the other gases. If we therefore adopt this value of $\dfrac{1}{a}$ we obtain

$$a = 273.$$

In consequence of equation (I.) we can treat any one of the three magnitudes p, v, and t—for example, p—as a function of the two others, v and t. These latter then are the independent variables by which the condition of the gas is fixed. We shall now seek to determine how the magnitudes which relate to the quantities of heat depend on these two variables.

If any body changes its volume, mechanical work will in general be either produced or expended. It is, however, in most cases impossible to determine this exactly, since besides the *external* work there is generally an unknown amount of *internal* work done. To avoid this difficulty, Carnot employed the ingenious method already referred to of allowing the body to undergo its various changes in succession, which are so arranged that it returns at last exactly to its original condition. In this case, if *internal* work is done in some of the changes, it is exactly compensated for in the others, and we may be sure that the *external* work, which remains over after the changes are completed, is all the work that has been done. Clapeyron has represented this process graphically in a very clear way, and we shall follow his presentation now for the permanent gases, with a slight alteration rendered necessary by our principle.

In the figure, let the abscissa oe represent the volume and the ordinate ea the pressure on a unit weight of gas, in a condition in which its temperature $= t$. We assume that the gas is contained in an extensible envelope, which, however, cannot exchange heat with it. If, now, it is allowed to expand in this envelope, its temperature would fall if no heat were imparted to it. To avoid this, let it be put in contact, during its expansion, with a body, *A*, which is kept at the constant temperature *t*, and which imparts just so much heat to the gas that its temperature also remains equal to *t*. During this expansion at constant temperature, its pressure diminishes according to the M. law, and may be represented by the ordinate of a curve, *ab*, which is a portion of an equilateral hyperbola. When the volume of the gas has increased in this way from *oe* to *of*, the body *A* is removed, and the expansion is allowed to continue without the introduction of more heat. The temperature will then fall, and the pressure diminish more rapidly than before. The law which is followed in this part of the process may be represented by the curve *bc*. When the volume of the gas has increased in this way from *of* to *og*, and its temperature has fallen from *t* to *r*, we begin to compress it, in order to restore it again to its original volume *oe*. If it were left to itself its temperature would again rise. This, however, we do not permit, but bring it in contact with a body, *B*, at the constant temperature *r*, to which it at once gives up the heat that is produced, so that it keeps the temperature *r* ; and while it is in contact with this body we compress it so far (by the amount *gh*) that the remaining compression *he* is exactly sufficient to raise its temperature from *r* to *t*, if during this last compression it gives up no heat. During the former compression the pressure increases according to the M. law, and is represented by the portion *cd* of an equilateral hyperbola. During the latter, on the other hand, the increase is more rapid and is represented by the curve *da*. This curve must end exactly at *a*, for since at the end of the operation the volume and temperature have again their original values, the same must be true of the pressure also, which is a function of them both. The gas is

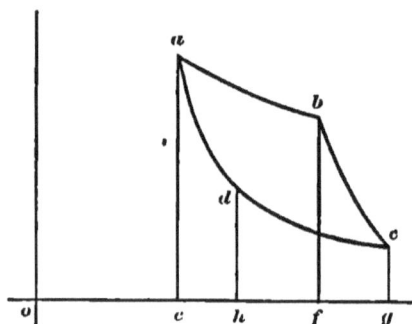

Fig. 1

74

therefore in the same condition again as it was at the beginning.

Now, to determine the work produced by these changes, for the reasons already given, we need to direct our attention only to the *external* work. During the expansion the gas *does* work, which is determined by the integral of the product of the differential of the volume into the corresponding pressure, and is therefore represented geometrically by the quadrilaterals *eabf* and *fbcg*. During the compression, on the other hand, work is *expended*, which is represented similarly by the quadrilaterals *gcdh* and *hdae*. The excess of the former quantity of work over the latter is to be looked on as the whole work produced during the changes, and this is represented by the quadrilateral *abcd*.

If the process above described is carried out in the reverse order, the same magnitude, *abcd*, is obtained as the excess of the work *expended* over the work *done*.

In order to make an analytical application of the method just described, we will assume that all the changes which the gas undergoes are *infinitely small*. We may then treat the curves obtained as straight lines, as they are represented in the accompanying figure. We may also, in determining the area of the quadrilateral *abcd*, consider it a parallelogram, since the error arising therefrom can only be a quantity of the *third* order, while the area itself is a quantity of the *second* order. On this assumption, as may easily be seen, the area may be represented by the product

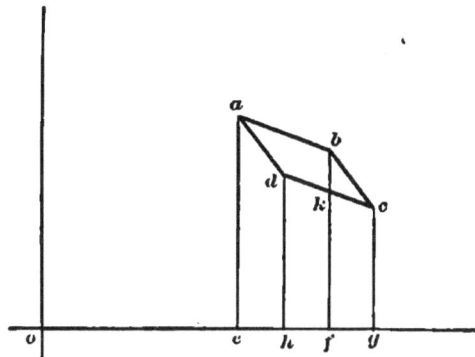

Fig. 2

ef.bk, if *k* is the point in which the ordinate *bf* cuts the lower side of the quadrilateral. The magnitude *bk* is the increase of the pressure, while the gas at the constant volume *of* has its temperature raised from τ to *t*—that is, by the differential $t-\tau=dt$. This magnitude may be at once expressed by the aid of equation (I.) in terms of *v* and *t*, and is

$$dp = \frac{Rdt}{v}.$$

If, further, we denote the increase of volume ef by dv, we obtain the area of the quadrilateral, and so, also,

(1) $$\textit{The work done} = \frac{Rdvdt}{v}.$$

We must now determine the heat consumed in these changes. The quantity of heat which must be communicated to a gas, while it is brought from any former condition in a definite way to that condition in which its volume $= v$ and its temperature $= t$, may be called Q, and the changes of volume in the above process, which must here be considered separately, may be represented as follows: ef by dv, hg by $d'v$, eh by δv, and fy by $\delta'v$. During an expansion from the volume $oe = v$ to the volume $of = v + dv$ at the constant temperature t, the gas must receive the quantity of heat

$$\left(\frac{dQ}{dv}\right)dv,$$

and correspondingly, during an expansion from $oh = v + \delta v$ to $og = v + \delta v + d'v$ at the temperature $t - dt$, the quantity of heat,

$$\left[\left(\frac{dQ}{dv} + \frac{d}{dv}\left(\frac{dQ}{dv}\right)\delta v - \frac{d}{dt}\left(\frac{dQ}{dv}\right)dt\right)\right]d'v.$$

In the case before us this latter quantity must be taken as negative in the calculation, because the real process was a compression instead of the expansion assumed. During the expansion from of to og and the compression from oh to oe, the gas has neither gained nor lost heat, and hence the quantity of heat which the gas has received in excess of that which it has given up—that is, the *heat consumed*

(2) $$= \left(\frac{dQ}{dv}\right)dv - \left[\left(\frac{dQ}{dv}\right) + \frac{d}{dv}\left(\frac{dQ}{dv}\right)\delta v - \frac{d}{dt}\left(\frac{dQ}{dv}\right)dt\right]d'v.$$

The magnitudes δv and $d'v$ must be eliminated from this expression. For this purpose we have first, immediately from the inspection of the figure, the following equation :

$$dv + \delta'v = \delta v + d'v.$$

From the condition that during the compression from oh to oe, and therefore also conversely during an expansion from oe to oh occurring under the same conditions, and similarly during the expansion from *of* to *og*, both of which occasion a fall of temperature by the amount dt, the gas neither receives nor gives up heat, we obtain the equations

$$\left(\frac{dQ}{dv}\right)\delta v - \left(\frac{dQ}{dt}\right)dt = 0,$$

$$\left[\left(\frac{dQ}{dv}\right) + \frac{d}{dv}\left(\frac{dQ}{dv}\right)dv\right]\delta'v - \left[\left(\frac{dQ}{dt}\right) + \frac{d}{dv}\left(\frac{dQ}{dt}\right)dv\right]dt = 0.$$

Eliminating from these three equations and equation (2) the three magnitudes $d'v$, δv, and $\delta'v$, and also neglecting in the development those terms which, in respect of the differentials, are of a higher order than the second, we obtain

(3)   *The heat consumed* $= \left[\dfrac{d}{dt}\left(\dfrac{dQ}{dv}\right) - \dfrac{d}{dv}\left(\dfrac{dQ}{dt}\right)\right]dvdt.$

If we now return to our principle, that to produce a certain amount of work the expenditure of a proportional quantity of heat is necessary, we can establish the formula

(4) $\dfrac{\textit{The heat consumed}}{\textit{The work done}} = A,$

where *A is a constant, which denotes the heat equivalent for the unit of work.* The expressions (1) and (3) substituted in this equation give

$$\frac{\left[\dfrac{d}{dt}\left(\dfrac{dQ}{dv}\right) - \dfrac{d}{dv}\left(\dfrac{dQ}{dt}\right)\right]dvdt}{\dfrac{R \cdot dvdt}{v}} = A,$$

or

(II.) $\dfrac{d}{dt}\left(\dfrac{dQ}{dv}\right) - \dfrac{d}{dv}\left(\dfrac{dQ}{dt}\right) = \dfrac{AR}{v}.$

We may consider this equation as the analytical expression of our fundamental principle applied to the case of permanent gases. It shows that Q cannot be a function of v and t, if these variables are independent of each other. For if it were, then by the well-known law of the differential calculus, that if a function of two variables is differentiated with respect to both of them, the order of differentiation is indifferent, the right-hand side of the equation should be equal to zero.

The equation may also be brought into the form of a *complete* differential equation,

· (II.a) $$dQ = dU + A . R\frac{a+t}{v}dv,$$

in which U is an arbitrary function of v and t. This differential equation is naturally not integrable, but becomes so only if a second relation is given between the variables, by which t may be treated as a function of v. The reason for this is found in the last term, and this corresponds exactly to the *external* work done during the change, since the differential of this work is pdv, from which we obtain

$$\frac{R(a+t)}{v}dv,$$

if we eliminate p by means of (I.).

We have thus obtained from equation (II.a) what was introduced before as an immediate consequence of our principle, that the total amount of heat received by the gas during a change of volume and temperature can be separated into two parts, one of which, U, which comprises the *free* heat that has entered and the heat *consumed* in doing *internal* work, if any such work has been done, has the properties which are commonly assigned to the total heat, of being a function of v and t, and of being therefore fully determined by the initial and final conditions of the gas, between which the transformation has taken place ; while the other part, which comprises the heat *consumed* in doing *external* work, is dependent not only on the terminal conditions, but on the whole course of the changes between these conditions.

Before we undertake to prepare this equation for further conclusions, we shall develop the analytical expression of our fundamental principle for the case of vapors at their maximum density.

In this case we have no right to apply the M. and G. law, and so must restrict ourselves to the principle alone. In order to obtain an equation from it, we again use the method given by Carnot and graphically presented by Clapeyron, with a slight modification. Consider a liquid contained in a vessel impenetrable ·by heat, of which, however, only a part is filled by the liquid, while the rest is left free for the vapor, which is at the maximum density corresponding to its temperature, t. The total volume of both liquid and vapor is represented in the accompanying figure by the abscissa *oe,* and the pressure of the vapor by the ordinate *ea*. Let the vessel now yield to the pressure and enlarge in volume while the liquid and vapor are in contact with a body, *A*, at the constant temperature *t*. As the volume increases, more liquid evaporates, but the heat which thus becomes latent is supplied from the body *A*, so that the temperature, and so also the pressure, of the vapor remain unchanged. If in this way the total volume is increased from *oe* to *of*, an amount of external work is done which is represented by the rectangle *eabf*. Now remove the body *A* and let the vessel increase in volume still further, while heat can neither enter nor leave it. In this process the vapor already present will expand, and also new vapor will be produced, and in consequence the temperature will fall and the pressure diminish. Let this process go on until the temperature has changed from *t* to *τ*, and the volume has become *og*. If the fall of pressure during this expansion is represented by the curve *bc*, the external work done in the process = *fbcg*.

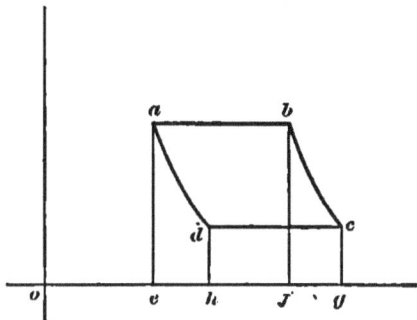

Now diminish the volume of the vessel, in order to bring the liquid with its vapor back to its original total volume, *oe*; and let this compression take place, in part, in contact with the body *B* at the temperature *τ*, into which body all the heat set free by the condensation of the vapor will pass, so that the temperature remains constant and = *τ*, in part without this body, so that the temperature rises. Let the operation be so managed that the first part of the compression is carried out only so far (to *oh*) that the volume *he* still remaining is exactly such that compression through it will raise the temperature from *τ* to *t* again. During the former diminution of volume the pressure remains invariable, = *gc*, and the external work employed is equal to the rectangle *gcdh*. During the latter diminution of volume the pressure increases and is represented by the curve *da*, which must end exactly at the point *a*, since the original pressure, *ea*, must correspond to the original temperature, *t*. The work employed in this last operation is = *hdae*. At the

79

end of the operation the liquid and vapor are again in the same condition as at the beginning, so that the excess of the *external* work done over that employed is also the *total* work done. It is represented by the quadrilateral *abcd*, and its area must also be set equal to the *heat consumed* during the same time.

For our purposes we again assume that the changes just described are infinitely small, and on this assumption represent the whole process by the accompanying figure, in which the curves *ad* and *bc* which occur in Fig. 3 have become straight lines. So far as the area of the quadrilateral *abcd* is concerned, it may again be considered a parallelogram, and may be represented by the product *ef.bk*. If, now, the pressure of the vapor at the temperature *t* and at its maximum tension is represented by *p*, and if the temperature difference $t - \tau$ is represented by *dt*, we

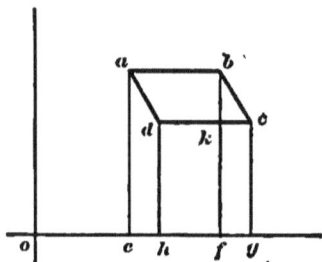

Fig. 4

have

$$bk = \frac{dp}{dt}\, dt.$$

The line *ef* represents the increase of volume, which occurs in consequence of the passage of a certain quantity of liquid, which may be called *dm*, over into vapor. Representing now the volume of a unit weight of the vapor at its maximum density at the temperature *t* by *s*, and the volume of the same quantity of liquid at the temperature *t* by σ, we have evidently

$$ef = (s - \sigma)\, dm,$$

and consequently the area of the quadrilateral, or

(5)   *The work done* $= (s - \sigma)\dfrac{dp}{dt} dm dt.$

In order to represent the quantities of heat concerned, we will introduce the following symbols. The quantity of heat which becomes latent when a unit weight of the liquid evaporates at the temperature *t* and under the corresponding pressure, is called r, and the specific heat of the liquid is called *c*. Both of these quantities, as well as also *s*, σ, and $\dfrac{dp}{dt}$, are to be considered functions of *t*. Finally, let us denote by *hdt* the quantity of heat which must be imparted to a unit weight of the vapor if its temperature is raised from t to $t + dt$, while it is so compressed that it is again at the maximum density for this temperature without the precipitation of any part of it. The quantity h is likewise a function of t. It will, for the present, be left undetermined whether it has a positive or negative value.

If we now denote by μ the mass of liquid originally present in the vessel, and by m the mass of vapor, and further by dm the mass which evaporates during the expansion from oe to of, and by $d'm$ the mass which condenses during the compression from og to oh, the heat which becomes latent in the first operation and is taken from the body A is

$$rdm,$$

and that which is set free in the second operation and is given to the body B is

$$(r - \frac{dr}{dt}dt)d'm.$$

In the other expansion and in the other compression heat is neither gained nor lost, so that, at the end of the process,

(6) $\quad$ *The heat consumed* $= rdm - (r - \frac{dr}{dt}dt)d'm.$

In this expression the differential $d'm$ must be replaced by dm and dt. For this purpose we make use of the conditions under which the second expansion and the second compression occurred. The mass of vapor, which condenses during the compression from oh to oe, and which would be evolved by the corresponding expansion from oe to oh, may be represented by δm, and that which is evolved by the expansion from of to og by $\delta'm$. We then have at once, since at the end of the process the same mass of liquid μ and the same mass of vapor m must be present as at the beginning, the equation

$$dm + \delta'm = d'm + \delta m.$$

Further, we obtain for the expansion from oe to oh, since in it the temperature of the mass of liquid μ and the mass of vapor m must be lowered by dt without the emission of heat, the equation

$$r\delta m - \mu.cdt - m.hdt = 0;$$

and similarly for the expansion from of to og, by substituting $\mu - dm$ and $m + dm$ for μ and m, and $\delta'm$ for δm,

$$r\delta'm - (\mu - dm)cdt - (m + dm)hdt = 0.$$

If from these three equations and (6) we eliminate the magnitudes $d'm$, δm, and $\delta'm$, and reject terms of higher order than the second, we have

(7)   *The heat consumed* $=\left(\dfrac{dr}{dt}+c-h\right)dmdt.$

The formulas (7) and (5) must now be connected in the same way as that used in the case of the permanent gases, that is,

$$\frac{\left(\dfrac{dr}{dt}+c-h\right)dmdt}{(s-\sigma)\dfrac{dp}{dt}\,dmdt}=A,$$

and we obtain as the analytical expression of the fundamental principle in the case of vapors at their maximum density the equation

(III.) $\dfrac{dr}{dt}+c-h=A(s-\sigma)\dfrac{dp}{dt}.$

If, instead of using our principle, we adopt the assumption that the quantity of heat is *constant*, we must replace (III.), as appears from (7), by

(8) $\dfrac{dr}{dt}+c-h=0.$

This equation has been used, if not exactly in the same form, at least in its general sense, to obtain a value for the magnitude h. So long as Watt's law is considered true for water, that the sum of the free and latent heats of a quantity of vapor at its maximum density is equal for all temperatures, and that therefore

$$\frac{dr}{dt}+c=0,$$

it must be concluded that for this liquid $h=0$. This conclusion has, in fact, often been stated as correct, in that it has been said that if a quantity of vapor is at its maximum density, and then compressed or expanded in a vessel impermeable by heat, it remains at its maximum density. But since Regnault[*] has corrected Watt's law by substituting for it the approximate relation

$$\frac{dr}{dt}+c=0.305,$$

the equation (8) gives for h the value 0.305. It would therefore follow that the quantity of vapor formerly considered in

---

the vessel impermeable by heat would be partly condensed by compression, and on expansion would not remain at the maximum density, since its temperature would not fall in a way to correspond to the diminution of pressure.

It is entirely different if we replace equation (8) by (III.). The expression on the right-hand side is, from its nature, always positive, and it therefore follows that h must be less than 0.305. It will subsequently appear that the value of this expression is so great that h is negative. We must therefore conclude that the quantity of vapor before mentioned is partly condensed, not by *compression*, but by *expansion*, and that by compression its temperature rises at a greater rate than the density increases, so that it does not remain at its maximum density.

It must be admitted that this result is exactly opposed to the common view already referred to; yet I do not believe that it is contradicted by any experimental fact. Indeed, it is more consistent than the former view with the behavior of steam as observed by Pambour. Pambour* found that the steam which issues from a locomotive after it has done its work always has the temperature at which the tension, observed at the same time, is a maximum. From this it follows either that $h=0$, as it was once thought to be, because this assumption agreed with Watt's law, accepted as probably true, or that h is *negative*. For if h were *positive*, the temperature of the vapor, when released, would be too high in comparison with its tension, and that could not have escaped Pambour's notice. If, on the other hand, h is *negative*, according to our former statement, there can never arise from this cause too low a temperature, but a part of the steam must become liquid, so as to maintain the rest at the proper temperature. This part need not be great, since a small quantity of vapor sets free on condensation a relatively large quantity of heat, and the water formed will probably be carried on mechanically by the rest of the steam, and will in such researches pass unnoticed, the more likely as it might be thought, if it were to be observed, that it was water from the boiler carried out mechanically.

The results thus far obtained have been deduced from the fundamental principle without any further hypothesis. The equation (II.a) obtained for permanent gases may, however, be

* *Traité des Locomotives*, second edition, and *Théorie des Machines à Vapeur*, second edition.

made much more fruitful by the help of an obvious subsidiary hypothesis. The gases show in their various relations, especially in the relation expressed by the M. and G. law between volume, pressure, and temperature, so great a regularity of behavior that we are naturally led to take the view that the mutual attraction of the particles, which acts within solid and liquid bodies, no longer acts in gases, so that while in the case of other bodies the heat which produces expansion must overcome not only the external pressure but the internal attraction as well, in the case of gases it has to do only with the external pressure. If this is the case, then during the expansion of a gas only so much heat becomes *latent* as is used in doing *external* work. There is, further, no reason to think that a gas, if it expands at constant temperature, contains more *free* heat than before. If this be admitted, we have the law : *a permanent gas, when expanded at constant temperature, takes up only so much heat as is consumed in doing external work during the expansion.* This law is probably true for any gas with the same degree of exactness as that attained by the M. and G. law applied to it.

From this it follows at once that

(9)
$$\left(\frac{dQ}{dv}\right)=A.R\frac{a+t}{v},$$

since, as already noticed, $R\frac{a+t}{v}dv$ represents the external work done during the expansion dv. It follows that the function U which occurs in (II.*a*) does not contain v, and the equation therefore takes the form

(II.*b*)
$$dQ=cdt+AR\frac{a+t}{v}dv,$$

where c can be a function of t only. It is even probable that this magnitude c, which represents the specific heat of the gas at constant volume, is a constant.

Now in order to apply this equation to special cases, we must introduce the relation between the variables Q, t, and v, which is obtained from the conditions of each separate case, into the equation, and so make it integrable. We shall here consider only a few simple examples of this sort, which are either interesting in themselves or become so by comparison with other theorems already announced.

We may first obtain the specific heats of the gas at constant volume and at constant pressure if in (II.b) we set $v=$const., and $p=$const. In the former case, $dv=0$, and (II.b) becomes

(10) $$\frac{dQ}{dt}=c.$$

In the latter case, we obtain from the condition $p=$const., by the help of equation (I.),

$$dv=\frac{Rdt}{p},$$

or

$$\frac{dv}{v}=\frac{dt}{a+t};$$

and this, substituted in (II.b), gives

(10a) $$\frac{dQ}{dt}=c'=c+AR,$$

if we denote by c' the specific heat at constant pressure.

It appears, therefore, that the *difference of the two specific heats of any gas is a constant magnitude*, AR. This magnitude also involves a simple relation among the different gases.

The complete expression for R is $\frac{p_0 v_0}{a+t_0}$, where p_0, v_0, and t_0 are any three corresponding values of p, v, and t for a unit of weight of the gas considered, and it therefore follows, as has already been mentioned in connection with the adoption of equation (I.), that R is inversely proportional to the specific gravity of the gas, and hence also that the same statement must hold for the difference $c'-c=AR$, since A is the same for all gases.

If we reckon the specific heat of the gas, not with respect to the unit of *weight*, but, as is more convenient, with respect to the unit of *volume*, we need only divide c and c' by v_0, if the volumes are taken at the temperature t_0 and pressure p_0. Designating these quotients by γ and γ', we obtain

(11) $$\gamma'-\gamma=\frac{A.R}{v_0}=A\frac{p_0}{a+t_0}.$$

In this last quantity nothing appears which is dependent on the particular nature of the gas, and *the difference of the specific heats referred to the unit of volume is therefore the same for all gases.*

This law was deduced by Clapeyron from Carnot's theory,

though the constancy of the difference $c'-c$, which we have deduced before, is not found in his work, where the expression given for it still has the form of a function of the temperature.

If we divide equation (11) on both sides by γ, we have

(12)
$$k-1=\frac{A}{\gamma}\cdot\frac{p_0}{a+t_0},$$

in which k, for the sake of brevity, is used for the quotient $\dfrac{\gamma'}{\gamma}$,

or, what amounts to the same thing, for the quotient $\dfrac{c'}{c}$. This quantity has acquired special importance in science from the theoretical discussion by Laplace of the propagation of sound in air. *The excess of this quotient over unity is therefore, for the different gases, inversely proportional to the specific heats of the same at constant volume, if these are referred to the unit of volume.* This law has, in fact, been found by Dulong from experiment[*] to be so nearly accurate that he has assumed it, in view of its theoretical probability, to be strictly accurate, and has therefore employed it, conversely, to calculate the specific heats of the different gases from the values of k determined by observation. It must, however, be remarked that the law is only theoretically justified when the M. and G. law holds, which is not the case with sufficient exactness for all the gases employed by Dulong.

If it is now assumed that the specific heat of gases at constant volume c is constant, which has been stated above to be very probable, the same follows for the specific heat at constant pressure, and consequently *the quotient of the two specific heats* $\dfrac{c'}{c}=k$ *is a constant.* This law, which Poisson has already assumed as correct on the strength of the experiments of Gay-Lussac and Welter, and has made the basis of his investigations on the tension and heat of gases,[†] is therefore in good agreement with our present theory, while it would not be possible on Carnot's theory as hitherto developed.

If in equation (II.b) we set $Q=$const., we obtain the following equation between v and t :

[*] *Ann. de Chim. et de Phys.*, xli., and Pogg. *Ann.*, xvi.
[†] *Traité de Mécanique*, second edition, vol. ii., p. 646.

(13) $$cdt + A. R \frac{a+t}{v} dv = 0,$$

which gives, if c is considered constant,

$$v \frac{A.R}{c} . (a+t) = \text{const.},$$

or, since from equation (10a), $\frac{AR}{c} = \frac{c'}{c} - 1 = k - 1$,

$$v^{k-1} (a+t) = \text{const.}$$

Hence we have, if v_0, t_0, and p_0 are three corresponding values of v, t, and p,

(14) $$\frac{a+t}{a+t_0} = \left(\frac{v_0}{v}\right)^{k-1}.$$

If we substitute in this relation the pressure p first for v and then for t by means of equation (I.), we obtain

(15) $$\left(\frac{a+t}{a+t_0}\right)^{k} = \left(\frac{p}{p_0}\right)^{k-1}$$

(16) $$\frac{p}{p_0} = \left(\frac{v_0}{v}\right)^{k.}$$

These are the relations which hold between volume, temperature, and pressure, if a quantity of gas is compressed or expanded within an envelope impermeable by heat. These equations agree precisely with those which have been developed by Poisson for the same case,* which depends upon the fact that he also treated k as a constant.

Finally, if we set $t = \text{const.}$ in equation (II.b), the first term on the right drops out, and there remains

(17) $$dQ = AR \frac{a+t}{v} dv,$$

from which we have

$$Q = AR (a+t) \log v + \text{const.},$$

or, if we denote by v_0, p_0, t_0, and Q_0 the values of v, p, t, and Q, which hold at the beginning of the change of volume,

(18) $$Q - Q_0 = AR(a + t_0) \log \frac{v}{v_0}.$$

From this follows the law also developed by Carnot: *If a gas changes its volume without changing its temperature, the quantities of heat evolved or absorbed are in arithmetical progression, while the volumes are in geometrical progression.*

---

* *Traité de Mécanique*, vol. ii., p. 647.

Further, if we substitute for R in (18) the complete expression $\frac{p_0 v_0}{a+t_0}$, we have

(19) $$Q - Q_0 = A\, p_0 v_0 \log \frac{v}{v_0}.$$

If now we apply this equation to the different gases, not by using equal *weights* of them, but such quantities as have at the outset equal volumes, v_0, it becomes in all its parts independent of the special nature of the gas, and agrees with the well-known law which Dulong proposed, guided by the above-mentioned simple relation of the magnitude $k-1$, *that all gases, if equal volumes of them are taken at the same temperature and under the same pressure, and if they are then compressed or expanded by an equal fraction of their volumes, either evolve or absorb an equal quantity of heat.* Equation (19) is, however, much more general. It states in addition, *that the quantity of heat is independent of the temperature at which the volume of the gas is altered,* if only the quantity of the gas employed is always determined so that the original volume v_0 is always the same at the different temperatures ; and it states further, that *if the original pressure is different in the different cases, the quantities of heat are proportional to it.*

II. CONSEQUENCES OF CARNOT'S PRINCIPLE IN CONNECTION WITH THE ONE ALREADY INTRODUCED

Carnot assumed, as has already been mentioned, that *the equivalent of the work done by heat is found in the mere transfer of heat from a hotter to a colder body, while the quantity of heat remains undiminished.*

The latter part of this assumption—namely, that the quantity of heat remains undiminished—contradicts our former principle, and must therefore be rejected if we are to retain that principle. On the other hand, the first part may still obtain in all its essentials. For though we do not need a special equivalent for the work done, since we have assumed as such an actual *consumption* of heat, it still may well be possible that such a transfer of heat occurs *at the same time* as the consumption of heat, and also stands in a definite relation to the work done. It becomes important, therefore, to consider whether this assumption, besides the mere possibility, has also a sufficient probability in its favor.

A transfer of heat from a hotter to a colder body always occurs in those cases in which work is done by heat, and in which
also the condition is fulfilled that the working substance is in
the same state at the end as at the beginning of the operation.
For example, we have seen, in the processes described above
and represented in Figs. 1 and 3, that the gas and the evaporating water took up heat from the body A as their volume increased, and gave it up to the body B as their volume diminished ; so that a certain quantity of heat was transferred from
A to B, and this was in fact much greater than that which we
assumed to be consumed, so that in the infinitely small changes,
which are represented in Figs. 2 and 4, the latter was an in-
finitesimal of the second order, while the former was one of
the first order. Yet, in order to establish a relation between
the heat transferred and the work done, a certain restric-
tion is necessary. For since a transfer of heat can take place
without mechanical effect if a hotter and a colder body are immediately in contact and heat passes from one to the other by
conduction, the way in which the transfer of a certain quantity
of heat between two bodies at the temperatures t and r can be
made to do the maximum of work is to so carry out the process, as was done in the above cases, that two bodies of different
temperatures never come in contact.

It is this *maximum* of work which must be compared with
the heat transferred. When this is done it appears that there
is in fact ground for asserting, with Carnot, that it depends
only on the quantity of the heat transferred and on the temperatures t and r of the two bodies A and B, but not on the nature
of the substance by means of which the work is done. This
maximum has, namely, the property that by *expending* it as
great a quantity of heat can be transferred from the cold body
B to the hot body A as passes from A to B when it is *produced*.
This may easily be seen, if we think of the whole process formerly described as carried out in the reverse order, so that, for
example, in the first case the gas first expands by itself, until
its temperature falls from t to r, is then expanded in connection
with B, is then compressed by itself until its temperature is
again t, and finally is compressed in connection with A. In this
case more work will be employed during the compression than
is produced during the expansion, so that on the whole there
is a loss of work, which is exactly as great as the gain of work in

the former process. Further, there will be just as much heat taken from the body B as was before given to it, and just as much given to the body A as was before taken from it, whence it follows not only that the same amount of heat is produced as was formerly consumed, but also that the heat which in the former process was transferred from A to B now passes from B to A.

If we now suppose that there are two substances of which the one can produce more work than the other by the transfer of a given amount of heat, or, what comes to the same thing, needs to transfer less heat from A to B to produce a given quantity of work, we may use these two substances alternately by producing work with one of them in the above process, and by expending work upon the other in the reverse process. At the end of the operations both bodies are in their original condition ; further, the work produced will have exactly counterbalanced the work done, and therefore, by our former principle, the quantity of heat can have neither increased nor diminished. The only change will occur in the *distribution* of the heat, since more heat will be transferred from B to A than from A to B, and so on the whole heat will be transferred from B to A. By repeating these two processes alternately it would be possible, without any expenditure of force or any other change, to transfer as much heat as we please from a *cold* to a *hot* body, and this is not in accord with the other relations of heat, since it always shows a tendency to equalize temperature differences and therefore to pass from *hotter* to *colder* bodies.

It seems, therefore, to be *theoretically* admissible to retain the first and the really essential part of Carnot's assumptions, and to apply it as a second principle in conjunction with the first ; and the correctness of this method is, as we shall soon see, established already in many cases by its *consequences.*

On this assumption we may express the maximum of work which can be produced by the transfer of a unit of heat from the body A at the temperature t to the body B at the temperature τ, as a function of t and τ. The value of this function must naturally be smaller as the difference $t-\tau$ is smaller, and when this is infinitely small ($=dt$) it must go over into the product of dt and a function of t only. For this latter case, with which we will concern ourselves for the present, the work may be expressed by the form $\frac{1}{C}.dt$, where C is a function of t only.

In order to apply this result to the permanent gases, we return to the process represented in Fig. 2. In that case the quantity of heat,

$$\left(\frac{dQ}{dv}\right)dv,$$

passed during the first expansion from A to the gas, and by the first compression the part of it expressed by

$$\left[\left(\frac{dQ}{dv}\right)+\frac{d}{dv}\left(\frac{dQ}{dv}\right)\delta v-\frac{d}{dt}\left(\frac{dQ}{dv}\right)dt\right]d'v,$$

or by

$$\left(\frac{dQ}{dv}\right)dv-\left[\frac{d}{dt}\left(\frac{dQ}{dv}\right)-\frac{d}{dv}\left(\frac{dQ}{dt}\right)\right]dvdt,$$

was given up to the body B. The latter magnitude is, therefore, the quantity of heat transferred. Since we may neglect the term of the second order with respect to the one of the first order, we retain simply

$$\left(\frac{dQ}{dv}\right)dv.$$

The work produced at the same time was

$$\frac{Rdv.dt}{v},$$

and we can thus at once form the equation

$$\frac{\dfrac{Rdv.dt}{v}}{\left(\dfrac{dQ}{dv}\right)dv}=\frac{1}{C}.dt,$$

or,

(IV.)
$$\left(\frac{dQ}{dv}\right)=\frac{RC}{v}.$$

If, in the second place, we make a similar application to the process represented in Fig. 4 relating to vaporization, we have for the quantity of heat carried from A to B

$$(r-\frac{dr}{dt}\,dt)d'm,$$

or
$$rdm-\left(\frac{dr}{dt}+c-h\right)dmdt,$$

for which, by neglecting the term of the second order, we may set simply

$$rdm.$$

The work produced was

$$(s-\sigma)\frac{dp}{dt}\,dm\,dt,$$

and we therefore get the equation

$$\frac{(s-\sigma)\frac{dp}{dt}.dm.dt,}{r\,dm}=\frac{1}{C}.dt,$$

or,

(V.) $$r=C.(s-\sigma)\frac{dp}{dt}.$$

These are the two analytical expressions of Carnot's principle, as they are given by Clapeyron in his memoir, in a somewhat different form. For vapors he stops with this equation (V.) and some immediate applications of it. For gases, on the other hand, he makes the equation (IV.) the basis of a more extended development. It is in this development that the partial disagreement appears between his results and ours.

We shall now connect these two equations with the results of the first principle, first considering equation (IV.) in connection with the consequences formerly deduced for the case of permanent gases.

If we restrict ourselves to that result which depends only on the fundamental principle—that is, to equation (II.a)—we can use equation (IV.) to further define the magnitude U, which appears there as an arbitrary function of v and t, and our equation becomes

(II.c) $$dQ=[B+R\Big(\frac{dC}{dt}-A\Big)\log v]\,dt+\frac{R.C}{v}dv,$$

where B is now an arbitrary function of t only.

If we also accept as correct the subsidiary hypothesis, then equation (IV.) is not necessary for the further definition of (II.a); since the same end is more completely attained by equation (9), which followed as an immediate consequence of this hypothesis in connection with the first principle. We gain, however, an opportunity to subject the results of the two principles to a comparative test. Equation (9) reads:

$$\Big(\frac{dQ}{dv}\Big)=\frac{R.A(a+t)}{v},$$

and if we compare this with (IV.), we see that they both express the same result, only the one in a more definite way than the other, since for the general temperature function denoted in (IV.) by C, the equation (9) gives the special expression $A\,(a+t)$.

To this striking agreement it may be added that equation (V.), in which also the function C appears, confirms the view that $A\,(a+t)$ is the correct expression for this function. This equation has been used by Clapeyron and Thomson to calculate the values of C for several temperatures. Clapeyron chose as the temperatures the boiling-points of ether, alcohol, water, and oil of turpentine, and by substituting in equation (V.) the values of $\dfrac{dp}{dt}$, s, and r for these liquids, determined by experiments at these boiling-points, he obtained for C the numbers contained in the second column of the table which follows. Thomson, on the other hand, considered *water vapor* only, but at different temperatures, and thence calculated the value of C for every degree between 0° and 230° Cent. For this purpose Regnault's series of observations have given him an admissible basis so far as the magnitudes $\dfrac{dp}{dt}$ and r are concerned; but the magnitude s is not so well known for other temperatures as for the boiling-point, and about this magnitude Thomson felt himself compelled to make an assumption, which he himself recognized as only approximately correct, and considered as a temporary aid, to be employed until more exact data are determined—namely, that water vapor at its maximum density follows the M. and G. law. The numbers which follow from his calculation for the same temperatures as those used by Clapeyron are given in the third column reduced to French units:

I

1	2	3
t IN CENT. DEGREES	C ACCORDING TO CLAPEYRON	C ACCORDING TO THOMSON
35°.5	0.733	0.728
78°.8	0.828	0.814
100°	0.897	0.855
156°.8	0.930	0.952

It appears that the values of C found in both cases increase slowly with the temperature, similarly to the values of $A\,(a+t)$. They are in the ratio of the numbers in the following rows :

$$1 : 1.13 : 1.22 : 1.27$$
$$1 : 1.12 : 1.17 : 1.31$$

and if we determine the ratios of the values of $A\,(a+t)$ corresponding to the same temperatures, we obtain

$$1 : 1.14 : 1.21 : 1.39.$$

This series of *relative* values diverges from the two others only so far as can be accounted for by the uncertainty of the data which underlie them. The same agreement will be shown later in connection with the determination of the constant A, in respect to the *absolute* values.

Such an agreement between results which are obtained from entirely different principles cannot be accidental ; it rather serves as a powerful confirmation of the two principles and the first subsidiary hypothesis annexed to them.

Returning again to the application of equations (IV.) and (V.), we may remark that the former, so far as relates to the permanent gases, has only served to confirm conclusions already obtained. In the consideration of vapors, and of all other substances to which Carnot's principle will be applied in the future, it furnishes, however, an essential improvement, in that it permits us to replace the function C, which recurs everywhere, by the definite expression $A\,(a+t)$.

By this substitution equation (V.) becomes

$$(\text{V.}a) \qquad r = A\,(a+t).(s-\sigma)\frac{dp}{dt},$$

and we therefore obtain for a vapor a simple relation between the temperature at which it is formed, the pressure, the volume, and the latent heat. This we can use in drawing further conclusions.

If the M. and G. law were accurate for vapors at their maximum density, we should have

$$(20) \qquad ps = R(a+t).$$

Eliminating the magnitude s from (V.a) by the use of this equation, and neglecting the magnitude σ, which vanishes in comparison with s if the temperature is not very high, we obtain

$$\frac{1}{p}\frac{dp}{dt} = \frac{r}{A\,R\,(a+t)^2}.$$

If we make the further assumption that r is constant, we obtain by integration, if p_1 denotes the tension of the vapor at 100°,

$$\log \frac{p_0}{p_1} = \frac{r\,(t-100)}{A.R\,(a+100)(a+t)},$$

or if we set $t-100=\tau$, $a+100=\alpha$, and $\dfrac{r}{A.R(a+100)}=\beta$,

(21) $\qquad\qquad \log \dfrac{p_0}{p_1} = \dfrac{\beta.\tau}{\alpha+\tau}.$

This equation cannot, of course, be accurate, since the two assumptions made in its development are not accurate; but since these, at least to a certain extent, approach the truth, the quantity $\dfrac{\beta.\tau}{\alpha+\tau}$ will roughly represent the value of the quantity $\log \dfrac{p_0}{p_1}$. We may explain in this way how it happens that this relation, if the constants α and β, instead of having values given them depending on their definitions, are considered as arbitrary, may serve as an empirical formula for the calculation of vapor tensions, without our being compelled to consider it as *fully* proved by theory, as is sometimes done.

The most immediate application of equation (V.a) is to *water vapor*, for which we have the largest collection of experimental data, in order to investigate *how far it departs, when at its maximum density, from the M. and G. law.* The magnitude of this departure cannot be unimportant, since carbonic acid and sulphurous acid, even at temperatures and tensions at which they are still far removed from their condensation points, show noticeable departures.

Equation (V.) may be put in the following form:

(22) $\qquad A p\,(s-\sigma)\dfrac{a}{a+t} = \dfrac{ar}{(a+t)^2\,\dfrac{1}{p}\dfrac{dp}{dt}}.$

The expression here found on the left-hand side would be very nearly constant, if the M. and G. law were applicable, since this law would give immediately, from (20),

$$A.ps\,\frac{a}{a+t} = A.Ra,$$

and $s-\sigma$ can be substituted for s in this equation with approximate accuracy. This expression is, therefore, especially suited to show clearly any departure from the M. and G. law, from the examination of its true values as they may be calculated from the expression on the right-hand side of (22). I have carried out this calculation for a series of temperatures, using for r and p the numbers given by Regnault.*

First with respect to the *latent heat:* Regnault states † that the quantity of heat λ, which must be imparted to a unit of weight of water, in order to heat it from $0°$ to $t°$ and then to evaporate it at that temperature, may be represented with tolerable accuracy by the formula:

(23) $$\lambda = 606.5 + 0.305\ t.$$

But now, from the significance of λ,

(23a) $$\lambda = r + \int_0^t c\,dt,$$

and for the magnitude c, the specific heat of water, which appears in this formula, Regnault has given the formula : ‡

(23b) $$c = 1 + 0.00004.t + 0.0000009.t^2.$$

By the help of these two equations we obtain for the latent heat from equation (23) the expression :

(24) $$r = 606.5 - 0.695.t - 0.00002.t^2 - 0.0000003.t^3. \ \S$$

Second, with respect to the pressure : in order to obtain from his numerous observations the most probable values, Regnault‖ made use of a graphic representation, by constructing curves, of which the abscissas represented the temperature and the ordinates the pressure p, and which are drawn in sections from $-33°$ to $+230°$. From $100°$ to $230°$ he has also

---

* *Mém. de l'Acad. de l'Inst. de France*, vol. xxi. (1847).

† Ibid., *Mém.* ix.; also Pogg. *Ann.*, Bd. 98.   ‡ Ibid., *Mém.* x.

§ In most of his investigations Regnault has not so much observed the heat which becomes *latent* by-evaporation of the vapor as that which becomes *free* by its condensation, and, therefore, since it has been shown above that, if the principle of the equivalence of heat and work is correct, the quantity of heat which a quantity of vapor gives up on condensation need not always be the same as that which it absorbs during its formation, the question may arise, whether such differences may not have entered in Regnault's experiments, so that the formula given for r would become inadmissible. I believe that we may answer this question in the negative, since Regnault so arranged his experiments that the condensation of the vapor occurred under the same pressure as its formation—that is, nearly under the pressure which corresponded as a maximum to the observed temperature, and in this case just as much heat must be evolved by condensation as is absorbed by evaporation.   ‖ Ibid., *Mém.* viii.

drawn a curve, of which the ordinates represent not p itself, but the logarithms of p. From this presentation the following values have been taken, which are to be considered as the immediate results of his observations, while the other *more complete* tables contained in the memoir were calculated from formulas, of which the choice and determination depended in the first instance upon these values:

II

t IN DEGREES CENTIGRADE ON THE AIR-THERMOMETER	p IN MILLI-METERS	t IN DEGREES CENTIGRADE ON THE AIR-THERMOMETER	p IN MILLIMETERS	
			FROM THE CURVE OF NUMBERS	FROM THE CURVE OF LOGARITHMS *
−20°	0.91	110°	1073.7	1073.3
−10	2.08	120	1489.0	1490.7
0	4.60	130	2029.0	2030.5
10	9.16	140	2713.0	2711.5
20	17.39	150	3572.0	3578.5
30	31.55	160	4647.0	4651.6
40	54.91	170	5960.0	5956.7
50	91.98	180	7545.0	7537.0
60	148.79	190	9428.0	9425.4
70	233.09	200	11660.0	11679.0
80	354.64	210	14308.0	14325.0
90	525.45	220	17390.0	17390.0
100	760.00	230	20915.0	20927.0

Now in order to carry out with these data the calculation in hand, I first determined from these tables the values of $\frac{1}{p}\cdot\frac{dp}{dt}$ for the temperatures −15°, −5°, 5°, 15°, etc., in the following way. Since the magnitude $\frac{1}{p}\cdot\frac{dp}{dt}$ only diminishes slowly as the temperature rises, I have considered as uniform the diminution in each interval of 10°, say from −20° to −10°, from −10° to 0°, etc., so that I could look on the value holding, for example, for 25° as the mean of all the values holding between 20° and 30°. On this assumption, since $\frac{1}{p}\cdot\frac{dp}{dt}=\frac{d\,(\log p)}{dt}$, I could use the formula:

---

* Instead of the *logarithms* obtained immediately from the curve and adopted by Regnault, the *numbers* corresponding to them are given, in order to facilitate comparison with the numbers in the next column.

$$\left(\frac{1}{p}\cdot\frac{dp}{dt}\right)_{25°} = \frac{\log p_{30°}-\log p_{20°}}{10},$$

or

(25)
$$\left(\frac{1}{p}\cdot\frac{dp}{dt}\right)_{25°} = \frac{\text{Log } p_{30°}-\text{Log } p_{20°}}{10\,.\,M},$$

in which Log indicates the Briggsian logarithms and M the modulus of this system. By help of these values of $\frac{1}{p}\cdot\frac{dp}{dt}$ and the values of r given by equation (24), and of the value 273 for a, the values which the expression on the right-hand side of (22), and so also the expression $Ap\,(s-\sigma)\,\dfrac{a}{a+t}$, take for the temperatures $-15°$, $-5°$, $5°$, etc., were calculated and are given in the accompanying table. For temperatures above $100°$ both series of numbers given for p are used separately, and the two results found in each case given opposite each other. The significance of the third and fourth columns will be indicated in the sequel.

III

1 t IN DEGREES CENTIGRADE ON THE AIR-THERMOMETER	$Ap\,(s-\sigma)\,\dfrac{a}{a+t}$		4 DIFFERENCES
	2 FROM THE OBSERVED VALUES	3 FROM EQUATION (27)	
−15	30.61	30.61	0.00
−5	29.21	30.54	+ 1.33
5	30.93	30.46	− 0.47
15	30.60	30.38	− 0.22
25	30.40	30.30	− 0.10
35	30.23	30.20	− 0.03
45	30.10	30.10	0.00
55	29.98	30.00	+ 0.02
65	29.88	29 88	0.00
75	29.76	29.76	0.00
85	29.65	29.63	− 0.02
95	29.49	29.48	− 0.01
105	29.47 29.50	29.33	− 0.14 − 0.17
115	29.16 29.02	29.17	+ 0.01 + 0.15
125	28.89 28.93	28.90	+ 0.10 + 0.06
135	28.88 29.01	28.80	− 0.08 − 0.21
145	28 65 28.40	28 60	− 0.05 + 0.20
155	28.16 28.25	28.38	+ 0.22 + 0.13
165	28.02 28.19	28.14	+ 0.12 − 0.05
175	27.84 27.90	27.89	+ 0.05 − 0.01
185	27.76 27.67	27.62	− 0.14 − 0.05
195	27.45 27.20	27.33	− 0.12 + 0.13
205	26.89 26.94	27.02	+ 0.13 + 0.08
215	26.56 26.79	26.68	+ 0.12 − 0.11
225	26.64 26.50	26.32	− 0.32 − 0.18

It appears at once from this table that $Ap\,(s-\sigma)\,\dfrac{a}{a+t}$ is not constant as it should be if the M. and G. law were applicable, but diminishes distinctly as the temperature rises. Between 35° and 90° this diminution appears to be very uniform. Under 35°, especially in the region of 0°, there appear noticeable irregularities, which, however, may be simply explained from the fact that in that region the pressure p and its differential coefficient $\dfrac{dp}{dt}$ are very small, and therefore small errors, which fall quite within the limits of the errors of observation, may become *relatively* important. It may be added that the curve by which the separate values of p are determined, as mentioned above, is not drawn in one stroke from — 35° to 100°, but, to economize space, is broken at 0°, so that at this temperature the progress of the curve cannot be determined so satisfactorily as it can within the separate portions below 0° and above 0°. From the way in which the differences occur in the foregoing table, it would seem that the value 4.60 mm. taken for p at 0° is a little too great, since if that were so the values of $Ap\,(s-\sigma)\,\dfrac{a}{a+t}$ for the temperatures just under 0° would come out too small, and for those just over 0° too large. Above 100° the values of this expression do not diminish so regularly as between 35° and 95°; and yet they show, at least *in general*, a corresponding progress; and especially if we use a graphic representation, we find that the curve, which within that interval almost exactly joins the successive points determined by the numbers contained in the table, may be produced beyond that interval even to 230° quite naturally, so that these points are evenly distributed on both sides of it.

Within the range of the table the progress of the curve can be represented with fair accuracy by an equation of the form

$$(26) \qquad Ap\,(s-\sigma)\,\frac{a}{a+t}=m-ne^{kt},$$

where e is the base of the natural logarithms, and m, n, and k are constants. If these constants are calculated from the values which the curve gives for 45°, 125°, and 205°, we obtain:

$$(26a) \qquad m=31.549,\ \ n=1.0486,\ \ k=0.007138,$$

and if for convenience we introduce the Briggsian logarithms, we obtain

(27) $\quad \text{Log} \left[31.549 - Ap\,(s - \sigma)\,\dfrac{a}{a + t} \right] = 0.0206 + 0.003100\,t.$

The numbers contained in the third column are calculated from this equation, and in the fourth are given the differences between these numbers and those in the second column.

From the foregoing we may easily deduce a formula by which we can more definitely determine the way in which the behavior of a vapor departs from the M. and G. law. By assuming this law to hold, and denoting by ps_0 the value of ps at $0°$, we would have from (20),

$$\frac{ps}{ps_0} = \frac{a + t}{a},$$

and would have, therefore, for the differential coefficient $\dfrac{d}{dt}\left(\dfrac{ps}{ps_0}\right)$ a constant quantity—namely, the well-known coefficient of expansion $\dfrac{1}{a} = 0.003665$. Instead of this we have from (26), if we simply replace $s - \sigma$ by s, the equation:

(28) $\quad \dfrac{ps}{ps_0} = \dfrac{m - ne^{kt}}{m - n} \cdot \dfrac{a + t}{a},$

and hence follows:

(29) $\quad \dfrac{d}{dt}\left(\dfrac{ps}{ps_0}\right) = \dfrac{1}{a} \cdot \dfrac{m - n\,[1 + k\,(a + t)]\,e^{kt}}{m - n}.$

The differential coefficient is, therefore, not a constant, but a function of the temperature which diminishes as the temperature increases. If we substitute the numerical values of m, n, and k, given in (26a), we obtain, among others, the following values for this function:

IV

t	$\dfrac{d}{dt}\left(\dfrac{ps}{ps_0}\right)$	t	$\dfrac{d}{dt}\left(\dfrac{ps}{ps_0}\right)$	t	$\dfrac{d}{dt}\left(\dfrac{ps}{ps_0}\right)$
Deg.		*Deg.*		*Deg.*	
0	0.00342	70	0.00307	140	0.00244
10	0.00338	80	0.00300	150	0.00231
20	0.00334	90	0.00293	160	0.00217
30	0.00320	100	0.00285	170	0.00203
40	0.00325	110	0.00276	180	0.00187
50	0.00319	120	0.00266	190	0.00168
60	0.00314	130	0.00256	200	0.00149

It appears from this table that at low temperatures the departures from the M. and G. law are only slight, but that at higher temperatures—for example, at 100°, and upwards—they can no longer be neglected.

It may appear at first sight remarkable that the values found for $\frac{d}{dt}\left(\frac{ps}{ps_0}\right)$ are *smaller* than 0.003665, since we know that in the case of gases, especially of those, like carbonic acid and sulphurous acid, which deviate most widely from the M. and G. law, the coefficient of expansion is not *smaller*, but *greater*, than that number. We are not, however, justified in making an immediate comparison between the differential coefficients which we have just determined and the coefficient of expansion in the ordinary sense of the words, which relate to the increase of volume *at constant pressure*, nor yet with the number obtained by keeping *the volume constant* during the heating process, and then observing the increase in the expansive force. We are dealing here with a third special case of the general differential coefficient $\frac{d}{dt}\left(\frac{ps}{ps_0}\right)$—namely, with that which arises when, as the heating goes on, the pressure increases in the same proportion as it does with water vapor when it is kept at its maximum density; and we must consider carbonic acid in these relations if we wish to institute a comparison.

Water vapor has a tension of 1^m at about 108°, and of 2^m at $129\frac{1}{2}°$. We will, therefore, examine the behavior of carbonic acid if it is heated by $21\frac{1}{2}°$, and if the pressure upon it is at the same time increased from 1^m to 2^m. According to Regnault [*] the coefficient of expansion of carbonic acid at the constant pressure 760^{mm} is 0.003710, and at the pressure 2520^{mm} is 0.003846. For a pressure of 1500^{mm} (the mean between 1^m and 2^m), if we consider the increase of the coefficient of expansion as proportional to the increase of pressure, we obtain the value 0.003767. If carbonic acid were heated at this mean pressure from 0° to $21\frac{1}{2}°$, the magnitude $\frac{pv}{pv_0}$ would increase from 1 to $1+0.003767\times21.5=1.08099$. Now from others of Regnault's researches [†] it is known that if carbonic acid, taken at a temperature near 0° under the pressure 1^m, is subjected to the pressure 1.98292^m, the magnitude ps decreases in the ratio of $1 : 0.99146$; so that for an increase of pressure from 1^m to 2^m there would be a decrease of this magnitude in the ratio of $1 : 0.99131$. If, now, both operations were performed at once— that is, the elevation of temperature from $0°$ to $21\frac{1}{2}°$ and the increase in pressure from 1^m to 2^m—the magnitude $\frac{pv}{pv_0}$ would increase nearly from 1 to $1.08099 \times 0.99131 = 1.071596$, and hence we obtain for the mean value of the differential coefficient $\frac{d}{dt}\left(\frac{pv}{pv_0}\right)$:

$$\frac{0.071596}{21.5} = 0.00333.$$

It appears, therefore, that in the case now under consideration, a value is obtained for carbonic acid which is less than 0.003665, and therefore a similar result for a vapor at *its maximum density* should not be considered at all improbable.

If, on the other hand, we were to determine the real coefficient of expansion of the vapor—that is, the number which expresses by how much a quantity of vapor expands if it is taken at a certain temperature at its maximum density, and then removed from the water and heated under constant pressure—we should certainly obtain a value which would be *greater*, and perhaps *considerably* greater, than 0.003665.

From equation (26) we easily obtain the *relative* volumes of a unit of weight of vapor at its maximum density for different temperatures, referred to the volume at some definite temperature. In order to calculate the *absolute* volumes from these with sufficient precision, we must know the value of the constant A with greater accuracy than is as yet the case.

The question now arises whether any one volume can be assigned with sufficient accuracy to permit its use as a starting-point in the calculation of the other absolute values from the relative values. Many investigations of the specific weight of water vapor have been carried out, the results of which, however, are not, in my opinion, conclusive for the case with which we are now dealing, in which the vapor is at its maximum density. The numbers which are ordinarily given, especially the one obtained by Gay-Lussac—0.6235—agree very well with the theoretical value obtained by assuming that 2 parts of hydrogen

and 1 part of oxygen combine to form 2 parts of water vapor—
that is, with the value

$$\frac{2 \times 0.06926 + 1.10563}{2} = 0.622.$$

These numbers, however, are obtained from observations which
were not carried out at temperatures at which the resulting
pressure was equal to the maximum expansive force, but at
higher temperatures. In this condition the vapor might nearly
conform to the M. and G. law, and the agreement with the
theoretical value may thus be explained. To pass from this
result to the condition of maximum density by the use of the
M. and G. law would contradict our previous conclusions, since
Table IV. shows too large a departure from this law, at the
temperatures at which the determination was made, to make
such a use of the law possible. Those experiments in which
the vapor was observed at its maximum density give for the
most part larger numbers, and Regnault has concluded * that
even at a temperature a little over 30°, in the case in which the
vapor is developed in *vacuum*, a sufficient agreement with the
theoretical value is reached only when the tension of the vapor
amounts to no more than 0.8 of that which corresponds to the
observed temperature as the maximum. A definite conclusion,
however, cannot be drawn from this observation, since it is
doubtful, as Regnault remarks, whether the departure is really
due to too great a specific weight of the vapor formed, or whether
a quantity of water remained condensed on the walls of the glass
globe. Other experiments, which were so executed that the
vapor did not form in vacuum but saturated a current of air,
gave results which were tolerably free from any irregularities,†
yet even these results, important as they are in other relations,
do not enable us to form any definite conclusions as to the behavior of vapor in a vacuum.

In this state of uncertainty the following considerations may
perhaps be of some service in filling the gap. Table IV. shows
that the vapor at its maximum density conforms more closely
to the M. and G. law as the temperature is lower, and it may
hence be concluded that the specific weight will approach the
theoretical value more nearly at lower than at higher temperatures. If therefore, for example, we assume the value 0.622

---

* *Ann. de Chim. et de Phys.*, III. Sér., t. xv., p. 148.   † Ibid., p. 158 ff.

as correct for 0° and then calculate the corresponding value d for higher temperatures by the help of the following equation deduced from (26),

$$(30) \qquad d = 0.622 \frac{m - n}{m - ne^{kt}},$$

we obtain much more probable values than if we were to adopt 0.622 as correct for all temperatures. The following table presents some of these values:

V

t	0°	50°	100°	150°	200°
d	0.622	0.631	0.645	0.666	0.698

Strictly speaking, we must go further than this. In Table III. we see that the values of $Ap\,(s - \sigma)\,\dfrac{a}{a+t}$, as the temperature falls, approach a limiting value, which is not reached even for the lowest temperatures of the table, and it is only for this limiting value that we have a right to assume the applicability of the M. and G. law and so set the specific weight equal to 0.622. The question therefore arises what this limiting value is. If we could consider the formula (26) as applicable for temperatures below $-15°$, we would have only to take the value which it approaches asymptotically, $m = 31.549$, and we could then replace equation (30) by the equation

$$(31) \qquad d = 0.622 \frac{m}{m - ne^{kt}}.$$

From this equation we obtain for the specific weight at 0° the value 0.643 instead of 0.622, and the other numbers of the preceding table must be increased in the same ratio. We are, however, not justified in so extended an application of formula (26), since it is only obtained empirically from the values given in Table III., and of these, those which relate to the lowest temperatures are rather uncertain. We must therefore, for the present, treat the limiting value of $A\,(s - \sigma)\,\dfrac{a}{a+t}$ as unknown, and content ourselves with such an approximation as the numbers in the preceding tables warrant. We may, however, conclude that these numbers are rather too small than too great.

If we combine equation (V.*a*) with equation (III.) deduced from the first principle, we may eliminate $A\,(s-\sigma)$, and obtain:

(32)
$$\frac{dr}{dt} + c - h = \frac{r}{a+t}.$$

By means of this equation we may determine the magnitude h, which has already been stated to be negative. If we set for c and r the expressions given in (23*b*) and (24), and for a the number 273, we obtain:

(33)
$$h = 0.305 - \frac{606. \; - 0.695\,t - 0.00002\,t^2 - 0.0000003\,t^3}{273 + t},$$

and hence obtain for h, among others, the values:

VI

t	0°	50°	100°	150°	200°
h	−1.916	−1.465	−1.133	−0.879	−0.676

In a way similar to that which we have followed in the case of water vapor, we might apply equation (V.*a*) to the vapors of other liquids also, and then compare the results obtained for these different liquids, as has been done with the numbers calculated by Clapeyron and contained in Table I. We shall not, however, go into these applications any further at present.

We must now endeavor to determine, at least approximately, the numerical value of the constant A, or, what is more useful, of the fraction $\frac{1}{A}$, that is, *the work equivalent of the unit of heat.*

For this purpose we can first use equation (10*a*) *for the permanent gases*, which amounts to the same thing as the method already employed by Mayer and Helmholtz. This equation is:
$$c' = c + A R,$$
and if we set for c the equivalent expression $\frac{c'}{k}$, we have:

(34)
$$\frac{1}{A} = \frac{k \cdot R}{(k-1)\,c'}.$$

The value commonly taken for c' for atmospheric air from the researches of De Laroche and Bérard is 0.267, and for k from the researches of Dulong is 1.421. Further, to determine $R = \frac{p_0\,v_0}{a+t_0}$, we know that the pressure of one atmosphere (760^{mm}) on a square meter is 10333 kilogrammes, and that the volume of one kilogramme of atmospheric air under that pressure and at the temperature of the freezing-point $= 0.7733$ cubic meters. Hence follows:

$$R = \frac{10333 \cdot 0.7733}{273} = 29.26,$$

and consequently

$$\frac{1}{A} = \frac{1.421 \cdot 29.26}{0.421 \cdot 0.267} = 370,$$

that is, by the expenditure of a unit of heat (that quantity of heat which will raise the temperature of 1 kilogramme of water from 0° to 1°) 370 kilogrammes can be lifted to the height of 1^m. Little confidence can be placed in this number, on account of the uncertainty of the numbers 0.267 and 1.421. Holtzmann gives as the limits, between which he is in doubt, 343 and 429.

We may further use the equation (V.a) developed for *vapors*. If we wish to apply it to *water* vapor, we can use the determinations given in the former part of our work, whose result is expressed in equation (26). If we choose in this equation the temperature 100°, for example, and set for p the corresponding pressure of 1 atmosphere $= 10333$ kilogrammes, we obtain:

(35) $$\frac{1}{A} = 257\,(s-\sigma).$$

If we now use Gay-Lussac's value of the specific weight of water vapor, 0.6235, we obtain $s = 1.699$, and hence,

$$\frac{1}{A} = 437.$$

Similar values are given by the use of the numbers contained in Table I., which Clapeyron and Thomson have calculated for C from equation (V.). For if we consider these as the values of $A\,(a+t)$ for the temperatures corresponding to them, we obtain for $\frac{1}{A}$ a set of values which lie between 416 and 462.

It has already been mentioned that the specific weight of water vapor given by Gay-Lussac is probably somewhat too small for the case where the vapor is at its maximum density. The same may be said of most of the specific weights which are ordinarily given for other vapors. We must therefore conclude that the values of $\frac{1}{A}$ calculated from them are for the most part a little too great. If we take for water vapor the number 0.645 given in Table V., from which $s=1.638$, we obtain

$$\frac{1}{A}=421.$$

This value is also perhaps a little, but probably not much, too great. We may therefore conclude, since this result should be given the preference over that obtained from atmospheric air, that *the work equivalent of the unit of heat is the lifting of something over 400 kilogrammes to the height of* 1^m.

We may now compare with this theoretical result those which Joule obtained in very different ways by direct observation. Joule obtained from the heat produced by magneto-electricity,

$$\frac{1}{A}=460\ ;*$$

from the quantity of heat which atmospheric air absorbs during its expansion,

$$\frac{1}{A}=438,\dagger$$

and as a mean of a large number of experiments, in which the heat produced by friction of water, of mercury, and of cast-iron, was observed,

$$\frac{1}{A}=425.\ddagger$$

The agreement of these three numbers, in spite of the difficulty of the experiments, leaves really no further doubt of the correctness of the fundamental principle of the equivalence of heat and work, and their agreement with the number 421 confirms in a similar way the correctness of Carnot's principle, in the form which it takes when combined with the first principle.

BIOGRAPHICAL SKETCH

RUDOLF JULIUS EMANUEL CLAUSIUS was born on January 2, 1822, at Cöslin, in Pomerania. He was educated at Berlin, and became Privat-docent in the University of Berlin and Instructor in Physics at the School of Artillery. In 1855 he was appointed to the Professorship of Physics in the Polytechnic School at Zürich, and in 1857 he was appointed to a similar

---

* *Phil. Mag.*, xxiii., p. 441. The number, given in English units, is reduced to French units.

† Ibid., xxvi., p. 381.　　　　　　　　　　　‡ Ibid., xxxv., p. 534.

position in the University of Zürich. In 1869 he was appointed Professor of Physics in the University of Bonn, where he remained until his death, on August 24, 1888.

Clausius was a prolific investigator and writer on physical subjects. The line of thought suggested by the discoveries in heat contained in the memoir given in this volume was followed out by him in a series of papers on the thermodynamic properties of bodies and on the general theory of thermodynamics. These papers were collected and published in a volume in 1864; and ten years later he recast these papers and others which had appeared after the collection was first published into a systematic treatise on the mechanical theory of heat. The concept of the entropy, which Clausius introduced and developed, is the most important single contribution made by him to science.

Clausius's investigations also extended into radiant heat, in connection with which he proved that radiance also conforms to the second law of thermodynamics. Clausius was the first to apply the doctrine of probabilities, in any systematic way, to the kinetic theory of gases; and by so doing he laid the foundations for the brilliant applications of that doctrine to the kinetic theories which have been made by Maxwell and Boltzmann. He also contributed something to the theory of electricity. His writings are characterized by simplicity of form and profundity of thought. They deal much with fundamental questions, but by such direct and simple methods that the ideas under discussion are rarely obscured by the difficulties of the analysis.

ON THE DYNAMICAL THEORY OF HEAT, WITH NUMERICAL RESULTS DEDUCED FROM MR. JOULE'S EQUIVALENT OF A THERMAL UNIT, AND M. REGNAULT'S OBSERVA- TIONS ON STEAM

BY

WILLIAM THOMSON (Lord Kelvin)

(*Transactions of the Royal Society of Edinburgh*, March, 1851 ; *Philosophical Magazine*, iv., 1852; *Mathematical and Physical Papers*, vol. i., p. 174)

CONTENTS

ON THE DYNAMICAL THEORY OF HEAT

BY

WILLIAM THOMSON

INTRODUCTORY NOTICE

1. SIR HUMPHRY DAVY, by his experiment of melting two pieces of ice by rubbing them together, established the following proposition: "The phenomena of repulsion are not dependent on a peculiar elastic fluid for their existence, or caloric does not exist." And he concludes that heat consists of a motion excited among the particles of bodies. "To distinguish this motion from others, and to signify the cause of our sensation of heat," and of the expansion or expansive pressure produced in matter by heat, "the name *repulsive* motion has been adopted."*

2. The dynamical theory of heat, thus established by Sir Humphry Davy, is extended to radiant heat by the discovery of phenomena, especially those of the polarization of radiant heat, which render it excessively probable that heat propagated through "vacant space," or through diathermanic substances, consists of waves of transverse vibrations in an all-pervading medium.

3. The recent discoveries made by Mayer and Joule,† of the

---

* From Davy's first work, entitled *An Essay on Heat, Light, and the Combinations of Light*, published in 1799, in " Contributions to Physical and Medical Knowledge, principally from the West of England, collected by Thomas Beddoes, M.D.," and republished in Dr. Davy's edition of his brother's collected works, vol. ii., Lond., 1836.

† In May, 1842, Mayer announced in the *Annalen* of Wöhler and Liebig, that he had raised the temperature of water from 12° to 13° Cent. by agitating it. In August, 1843, Joule announced to the British Association

generation of heat through the friction of fluids in motion, and by the magneto-electric excitation of galvanic currents, would either of them be sufficient to demonstrate the immateriality of heat; and would so afford, if required, a perfect confirmation of Sir Humphry Davy's views.

4. Considering it as thus established, that heat is not a substance, but a dynamical form of mechanical effect, we perceive that there must be an equivalence between mechanical work and heat, as between cause and effect. The first published statement of this principle appears to be in Mayer's *Bemerkungen über die Kräfte der unbelebten Natur,** which contains some correct views regarding the mutual convertibility of heat and mechanical effect, along with a false analogy between the approach of a weight to the earth and a diminution of the volume of a continuous substance, on which an attempt is founded to find numerically the mechanical equivalent of a given quantity of heat. In a paper published about fourteen months later, "On the Calorific Effects of Magneto-Electricity and the Mechanical Value of Heat,"† Mr. Joule, of Manchester, expresses very distinctly the consequences regarding the mutual convertibility of heat and mechanical effect which follow from the fact that heat is not a substance but a state of motion; and investigates on unquestionable principles the "absolute numerical relations," according to which heat is connected with mechanical power; verifying experimentally, that whenever heat is generated from purely mechanical action, and no other effect produced, whether it be by means of the friction of fluids or by the magneto-electric excitation of galvanic currents, the same quantity is generated by the same amount of work spent; and determining the actual amount of work, in foot-pounds, required to generate a unit of heat, which he calls "the mechanical equivalent of heat." Since the publica-

---

"That heat is evolved by the passage of water through narrow tubes;" and that he had "obtained one degree of heat per pound of water from a mechanical force capable of raising 770 pounds to the height of one foot;" and that heat is generated when work is spent in turning a magneto-electric machine, or an electro-magnetic engine. (See his paper "On the Calorific Effects of Magneto-Electricity, and on the Mechanical Value of Heat."—*Phil. Mag.*, vol. xxiii., 1843.)

* *Annalen* of Wöhler and Liebig, May, 1842.

† British Association, August, 1843; and *Phil. Mag.*, September, 1843.

tion of that paper, Mr. Joule has made numerous series of experiments for determining with as much accuracy as possible the mechanical equivalent of heat so defined, and has given accounts of them in various communications to the British Association, to the *Philosophical Magazine*, to the Royal Society, and to the French Institute.

5. Important contributions to the dynamical theory of heat have recently been made by Rankine and Clausius; who, by mathematical reasoning analogous to Carnot's on the motive power of heat, but founded on an axiom contrary to his fundamental axiom, have arrived at some remarkable conclusions. The researches of these authors have been published in the *Transactions* of this Society, and in Poggendorff's *Annalen*, during the past year; and they are more particularly referred to below in connection with corresponding parts of the investigations at present laid before the Royal Society.

6. The object of the present paper is threefold:

(1) To show what modifications of the conclusions arrived at by Carnot, and by others who have followed his peculiar mode of reasoning regarding the motive power of heat, must be made when the hypothesis of the dynamical theory, contrary as it is to Carnot's fundamental hypothesis, is adopted.

(2) To point out the significance in the dynamical theory, of the numerical results deduced from Regnault's observations on steam, and communicated about two years ago to the Society, with an account of Carnot's theory, by the author of the present paper; and to show that by taking these numbers (subject to correction when accurate experimental data regarding the density of saturated steam shall have been afforded), in connection with Joule's mechanical equivalent of a thermal unit, a complete theory of the motive power of heat, within the temperature limits of the experimental data, is obtained.

(3) To point out some remarkable relations connecting the physical properties of all substances, established by reasoning analogous to that of Carnot, but founded in part on the contrary principle of the dynamical theory.

PART I

Fundamental Principles in the Theory of the Motive Power of Heat

7. According to an obvious principle, first introduced, however, into the theory of the motive power of heat by Carnot, mechanical effect produced in any process cannot be said to have been derived from a purely thermal source, unless at the end of the process all the materials used are in precisely the same physical and mechanical circumstances as they were at the beginning. In some conceivable "thermo - dynamic engines," as, for instance, Faraday's floating magnet, or Barlow's "wheel and axle," made to rotate and perform work uniformly by means of a current continuously excited by heat communicated to two metals in contact, or the thermo-electric rotatory apparatus devised by Marsh, which has been actually constructed, this condition is fulfilled at every instant. On the other hand, in all thermo - dynamic engines, founded on electrical agency, in which discontinuous galvanic currents, or pieces of soft iron in a variable state of magnetization, are used, and in all engines founded on the alternate expansions and contractions of media, there are really alterations in the condition of materials ; but, in accordance with the principle stated above, these alterations must be strictly periodical. In any such engine the series of motions performed during a period, at the end of which the materials are restored to precisely the same condition as that in which they existed at the beginning, constitutes what will be called a complete cycle of its operations. Whenever in what follows, *the work done* or *the mechanical effect produced* by a thermo-dynamic engine is mentioned without qualification, it must be understood that the mechanical effect produced, either in a non-varying engine, or in a complete cycle, or any number of complete cycles of a periodical engine, is meant.

8. The *source of heat* will always be supposed to be a hot body at a given constant temperature put in contact with some part of the engine ; and when any part of the engine is to be kept from rising in temperature (which can only be done by drawing off whatever heat is deposited in it), this will be supposed to be done by putting a cold body, which will be

called the refrigerator, at a given constant temperature in contact with it.

9. The whole theory of the motive power of heat is founded on the two following propositions, due respectively to Joule, and to Carnot and Clausius.

PROP. I. (Joule).—When equal quantities of mechanical effect are produced by any means whatever from purely thermal sources, or lost in purely thermal effects, equal quantities of heat are put out of existence or are generated.

PROP. II. (Carnot and Clausius).—If an engine be such that, when it is worked backwards, the physical and mechanical agencies in every part of its motions are all reversed, it produces as much mechanical effect as can be produced by any thermo-dynamic engine, with the same temperatures of source and refrigerator, from a given quantity of heat.

10. The former proposition is shown to be included in the general "principle of mechanical effect," and is so established beyond all doubt by the following demonstration.

11. By whatever direct effect the heat gained or lost by a body in any conceivable circumstances is tested, the measurement of its quantity may always be founded on a determination of the quantity of some standard substance, which it or any equal quantity of heat could raise from one standard temperature to another; the test of equality between two quantities of heat being their capability of raising equal quantities of any substance from any temperature to the same higher temperature. Now, according to the dynamical theory of heat, the temperature of a substance can only be raised by working upon it in some way so as to produce increased thermal motions within it, besides effecting any modifications in the mutual distances or arrangements of its particles which may accompany a change of temperature. The work necessary to produce this total mechanical effect is of course proportional to the quantity of the substance raised from one standard temperature to another; and therefore when a body, or a group of bodies, or a machine, parts with or receives heat, there is in reality mechanical effect produced from it, or taken into it, to an extent precisely proportional to the quantity of heat which it emits or absorbs. But the work which any external forces do upon it, the work done by its own molecular forces, and the amount by which the half *vis viva* of the thermal motions of

all its parts is diminished, must together be equal to the mechanical effect produced from it : and, consequently, to the mechanical equivalent of the heat which it emits (which will be positive or negative, according as the sum of those terms is positive or negative). Now let there be either no molecular change or alteration of temperature in any part of the body, or, by a cycle of operations, let the temperature and physical condition be restored exactly to what they were at the beginning ; the second and third of the three parts of the work which it has to produce vanish ; and we conclude that the heat which it emits or absorbs will be the thermal equivalent of the work done upon it by external forces, or done by it against external forces ; which is the proposition to be proved.

12. The demonstration of the second proposition is founded on the following axiom : '

*It is impossible, by means of inanimate material agency, to derive mechanical effect from any portion of matter by cooling it below the temperature of the coldest of the surrounding objects.**

13. To demonstrate the second proposition, let A and B be two thermo-dynamic engines, of which B satisfies the conditions expressed in the enunciation ; and let, if possible, A derive more work from a given quantity of heat than B, when their sources and refrigerators are at the same temperatures, respectively. Then on account of the condition of complete *reversibility* in all its operations which it fulfils, B may be worked backwards, and made to restore any quantity of heat to its source, by the expenditure of the amount of work which, by its forward action, it would derive from the same quantity of heat. If, therefore, B be worked backwards, and made to restore to the source of A (which we may suppose to be adjustable to the engine B) as much heat as has been drawn from it during a certain period of the working of A, a smaller amount of work will be spent thus than was gained by the working of A. Hence, if such a series of operations of A forwards and of B backwards be continued, either alternately or simultaneously, there will result a continued production of work with-

---

* If this axiom be denied for all temperatures, it would have to be admitted that a self-acting machine might be set to work and produce mechanical effect by cooling the sea or earth, with no limit but the total loss of heat from the earth and sea, or, in reality, from the whole material world.

out any continued abstraction of heat from the source; and, by Prop. I., it follows that there must be more heat abstracted from the refrigerator by the working of B backwards than is deposited in it by A. Now it is obvious that A might be made to spend part of its work in working B backwards, and the whole might be made self-acting. Also, there being no heat either taken from or given to the source of the whole, all the surrounding bodies and space except the refrigerator might, without interfering with any of the conditions which have been assumed, be made of the same temperature as the source, whatever that may be. We should thus have a self-acting machine, capable of drawing heat constantly from a body surrounded by others at a higher temperature, and converting it into mechanical effect. But this is contrary to the axiom, and therefore we conclude that the hypothesis that A derives more mechanical effect from the same quantity of heat drawn from the source than B is false. Hence no engine whatever, with source and refrigerator at the same temperatures, can get more work from a given quantity of heat introduced than any engine which satisfies the condition of reversibility, which was to be proved.

14. This proposition was first enunciated by Carnot, being the expression of his criterion of a perfect thermo-dynamic engine.* He proved it by demonstrating that a negation of it would require the admission that there might be a self-acting machine constructed which would produce mechanical effect indefinitely, without any source either in heat or the consumption of materials, or any other physical agency; but this demonstration involves, fundamentally, the assumption that, in "a complete cycle of operations," the medium parts with exactly the same quantity of heat as it receives. A very strong expression of doubt regarding the truth of this assumption, as a universal principle, is given by Carnot himself;† and that it is false, where mechanical work is, on the whole, either gained or spent in the operations, may (as I have tried to show above) be considered to be perfectly certain. It must then be admitted that Carnot's original demonstration utterly fails, but we cannot infer that the proposition concluded is false. The truth of the conclusion appeared to me, indeed, so probable

---

* "Account of Carnot's Theory," § 13.   † Ibid., § 6.

that I took it in connection with Joule's principle, on account of which Carnot's demonstration of it fails, as the foundation of an investigation of the motive power of heat in air-engines or steam-engines through finite ranges of temperature, and obtained about a year ago results, of which the substance is given in the second part of the paper at present communicated to the Royal Society. It was not until the commencement of the present year that I found the demonstration given above, by which the truth of the proposition is established upon an axiom (§ 12) which I think will be generally admitted. It is with no wish to claim priority that I make these statements, as the merit of first establishing the proposition upon correct principles is entirely due to Clausius, who published his demonstration of it in the month of May last year, in the second part of his paper on the motive power of heat.* I may be allowed to add that I have given the demonstration exactly as it occurred to me before I knew that Clausius had either enunciated or demonstrated the proposition. The following is the axiom on which Clausius's demonstration is founded:

It is impossible for a self-acting machine, unaided by any external agency, to convey heat from one body to another at a higher temperature.

It is easily shown that, although this and the axiom I have used are different in form, either is a consequence of the other. The reasoning in each demonstration is strictly analogous to that which Carnot originally gave.

15. A complete theory of the motive power of heat would consist of the application of the two propositions demonstrated above to every possible method of producing mechanical effect from thermal agency.† As yet this has not been done for the electrical method, as far as regards the criterion of a perfect engine implied in the second proposition, and probably cannot be done without certain limitations; but the application of the first proposition has been very thoroughly investigated, and verified experimentally by Mr. Joule in his researches "On the

---

* Poggendorff's *Annalen*, referred to above.

† "There are at present known two, and only two, distinct ways in which mechanical effect can be obtained from heat. One of these is by the alterations of volume which bodies experience through the action of heat; the other is through the medium of electric agency."—"Account of Carnot's Theory," § 4. (*Transactions*, vol. xvi., part 5.)

Calorific Effects of Magneto-Electricity;" and on it is founded one of his ways of determining experimentally the mechanical equivalent of heat. Thus from his discovery of the laws of generation of heat in the galvanic circuit,* it follows that when mechanical work by means of a magneto - electric machine is the source of the galvanism, the heat generated in any given portion of the fixed part of the circuit is proportional to the whole work spent; and from his experimental demonstration that heat is developed in any moving part of the circuit at exactly the same rate as if it were at rest, and traversed by a current of the same strength, he is enabled to conclude :

(1) That heat may be created by working a magneto-electric machine.

(2) That if the current excited be not allowed to produce any other than thermal effects, the total quantity of heat produced is in all circumstances exactly proportional to the quantity of work spent.

16. Again, the admirable discovery of Peltier, that cold is produced by an electrical current passing from bismuth to antimony, is referred to by Joule,† as showing how it may be proved that, when an electrical current is continuously produced from a

---

* That, in a given fixed part of the circuit, the heat evolved in a given time is proportional to the square of the strength of the current, and for different fixed parts, with the same strength of current, the quantities of heat evolved in equal times are as the resistances. A paper by Mr. Joule, containing demonstrations of these laws, and of others on the relations of the chemical and thermal agencies concerned, was communicated to the Royal Society on the 17th of December, 1840, but was not published in the *Transactions*. (See abstract containing a statement of the laws quoted above, in the *Philosophical Magazine*, vol. xviii., p. 308.) It was published in the *Philosophical Magazine* in October, 1841 (vol. xix., p. 260).

† [Note of March 20, 1852, added in *Phil. Mag.* reprint. In the introduction to his paper "On the Calorific Effects of Magneto-Electricity," etc., *Phil. Mag.*, 1843.

I take this opportunity of mentioning that I have only recently become acquainted with Helmholtz's admirable treatise on the principle of mechanical effect (*Ueber die Erhaltung der Kraft*, von Dr. H. Helmholtz. Berlin. G. Reimer, 1847), having seen it for the first time on the 20th of January of this year; and that I should have had occasion to refer to it on this, and on numerous other points of the dynamical theory of heat, the mechanical theory of electrolysis, the theory of electro-magnetic induction, and the mechanical theory of thermo-electric currents, in various papers communicated to the Royal Society of Edinburgh, and to this Magazine, had I been acquainted with it in time.—W. T., March 20, 1852.]

purely thermal source, the quantities of heat evolved electrically in the different homogeneous parts of the circuit are only compensations for a loss from the junctions of the different metals, or that, when the effect of the current is entirely thermal, there must be just as much heat emitted from the parts not affected by the source as is taken from the source.

17. Lastly,* when a current produced by thermal agency is made to work an engine and produce mechanical effect, there will be less heat emitted from the parts of the circuit not affected by the source than is taken in from the source, by an amount precisely equivalent to the mechanical effect produced; since Joule demonstrates experimentally that a current from any kind of source driving an engine, produces in the engine just as much less heat than it would produce in a fixed wire exercising the same resistance as is equivalent to the mechanical effect produced by the engine.

18. The quality of thermal effects, resulting from equal causes through very different means, is beautifully illustrated

---

* This reasoning was suggested to me by the following passage contained in a letter which I received from Mr. Joule on the 8th of July, 1847. "In Peltier's experiment on cold produced at the bismuth and antimony solder, we have an instance of the conversion of heat into the mechanical force of the current," which must have been meant as an answer to a remark I had made, that no evidence could be adduced to show that heat is ever put out of existence. I now fully admit the force of that answer; but it would require a proof that there is more heat put out of existence at the heated soldering [or in this and other parts of the circuit] than is created at the cold soldering [and the remainder of the circuit, when a machine is driven by the current] to make the "evidence" be *experimental*. That this is the case I think is certain, because the statements of § 16 in the text are demonstrated consequences of the first fundamental proposition; but it is still to be remarked that neither in this nor in any other case of the production of mechanical effect from purely thermal agency, has the ceasing to exist of an equivalent quantity of heat been demonstrated otherwise than theoretically. It would be a very great step in the experimental illustration (or *verification*, for those who consider such to be necessary) of the dynamical theory of heat, to actually show in any one case a loss of heat; and it might be done by operating through a very considerable range of temperatures with a good air-engine or steam-engine, not allowed to waste its work in friction. As will be seen in Part II. of this paper, no experiment of any kind could show a considerable loss of heat without employing bodies differing considerably in temperature; for instance, a loss of as much as .098, or about one-tenth of the whole heat used, if the temperature of all the bodies used be between 0° and 30° Cent.

by the following statement, drawn from Mr. Joule's paper on magneto-electricity.*

Let there be three equal and similar galvanic batteries furnished with equal and similar electrodes; let A_1 and B_1 be the terminations of the electrodes (or wires connected with the two poles) of the first battery, A_2 and B_2 the terminations of the corresponding electrodes of the second, and A_3 and B_3 of the third battery. Let A_1 and B_1 be connected with the extremities of a long fixed wire; let A_2 and B_2 be connected with the "poles" of an electrolytic apparatus for the decomposition of water; and let A_3 and B_3 be connected with the *poles* (or *ports* as they might be called) of an electro-magnetic engine. Then if the length of the wire between A_1 and B_1, and the speed of the engine between A_3 and B_3, be so adjusted that the strength of the current (which for simplicity we may suppose to be continuous and perfectly uniform in each case) may be the same in the three circuits, there will be more heat given out in any time in the wire between A_1 and B_1 than in the electrolytic apparatus between A_2 and B_2, or the working engine between A_3 and B_3. But if the hydrogen were allowed to burn in the oxygen, within the electrolytic vessel, and the engine to waste all its work without producing any other than thermal effects (as it would do, for instance, if all its work were spent in continuously agitating a limited fluid mass), the total heat emitted would be precisely the same in each of these two pieces of apparatus as in the wire between A_1 and B_1. It is worthy of remark that these propositions are *rigorously* true, being demonstrable consequences of the fundamental principle of the dynamical theory of heat, which have been discovered by Joule, and illustrated and verified most copiously in his experimental researches.

19. Both the fundamental propositions may be applied in a perfectly rigorous manner to the second of the known methods of producing mechanical effect from thermal agency. This application of the first of the two fundamental propositions has already been published by Rankine and Clausius; and that of the second, as Clausius showed in his published paper, is sim-

---

* In this paper reference is made to his previous paper "On the Heat of Electrolysis" (published in vol. vii., part 2, of the second series of the Literary and Philosophical Society of Manchester) for experimental demonstration of those parts of the theory in which chemical action is concerned.

ply Carnot's unmodified investigation of the relation between the mechanical effect produced and the thermal circumstances from which it originates, in the case of an expansive engine working within an infinitely small range of temperatures. The simplest investigation of the consequences of the first proposition in this application, which has occurred to me, is the following, being merely the modification of an analytical expression of Carnot's axiom regarding the permanence of heat, which was given in my former paper,* required to make it express, not Carnot's axiom, but Joule's.

20. Let us suppose a mass† of any substance, occupying a volume v, under a pressure p uniform in all directions, and at a temperature t, to expand in volume to $v + dv$, and to rise in temperature to $t + dt$. The quantity of work which it will produce will be

$$pdv ;$$

and the quantity of heat which must be added to it to make its temperature rise during the expansion to $t + dt$ may be denoted by

$$Mdv + Ndt.$$

The mechanical equivalent of this is

$$J(Mdv + Ndt),$$

if J denote the mechanical equivalent of a unit of heat. Hence the mechanical measure of the total external effect produced in the circumstances is

$$(p - JM)\,dv - JNdt.$$

The total external effect, after any finite amount of expansion, accompanied by any continuous change of temperature, has taken place, will consequently be, in mechanical terms,

$$\int \{(p - JM)\,dv - JNdt\} ;$$

where we must suppose t to vary with v, so as to be the actual temperature of the medium at each instant, and the integration with reference to v must be performed between limits corresponding to the initial and final volumes. Now if, at any subsequent time, the volume and temperature of the medium become what they were at the beginning, however arbitrarily

---

* "Account of Carnot's Theory," foot-note on § 26.

† This may have parts consisting of different substances, or of the same substance in different states, provided the temperature of all be the same. See below, part iii., § 53–56.

they may have been made to vary in the period, the total external effect must, according to Prop. I., amount to nothing; and hence

$$(p - JM)\, dv - JN dt$$

must be the differential of a function of two independent variables, or we must have

$$\frac{d\,(p - JM)}{dt} = \frac{d\,(-JN)}{dv}, \qquad (1)$$

this being merely the analytical expression of the condition, that the preceding integral may vanish in every case in which the initial and final values of v and t are the same, respectively. Observing that J is an absolute constant, we may put the result into the form

$$\frac{dp}{dt} = J\left(\frac{dM}{dt} - \frac{dN}{dv}\right). \qquad (2)$$

This equation expresses, in a perfectly comprehensive manner, the application of the first fundamental proposition to the thermal and mechanical circumstances of any substance whatever, under uniform pressure in all directions, when subjected to any possible variations of temperature, volume, and pressure.

21. The corresponding application of the second fundamental proposition is completely expressed by the equation

$$\frac{dp}{dt} = \mu M, \qquad (3)$$

where μ denotes what is called "Carnot's function," a quantity which has an absolute value, the same for all substances for any given temperature, but which may vary with the temperature in a manner that can only be determined by experiment. To prove this proposition, it may be remarked in the first place that Prop. II. could not be true for every case in which the temperature of the refrigerator differs infinitely little from that of the source, without being true universally. Now, if a substance be allowed first to expand from v to $v + dv$, its temperature being kept constantly t; if, secondly, it be allowed to expand further, without either emitting or absorbing heat till its temperature goes down through an infinitely small range, to $t - \tau$; if, thirdly, it be compressed at the constant temperature $t - \tau$, so much (actually by an amount differing from dv by only an infinitely small quantity of the second order), that when, fourthly, the volume is further diminished to

r without the medium's being allowed to either emit or absorb heat, its temperature may be exactly t; it may be considered as constituting a thermo-dynamic engine which fulfils Carnot's condition of complete reversibility. Hence, by Prop. II., it must produce the same amount of work for the same quantity of heat absorbed in the first operation, as any other substance similarly operated upon through the same range of temperatures. But $\frac{dp}{dt} r . dv$ is obviously the whole work done in the complete cycle, and (by the definition of M in § 20) $M dv$ is the quantity of heat absorbed in the first operation. Hence the value of

$$\frac{\frac{dp}{dt} r . dv}{M dv}, \text{ or } \frac{dp}{dt} \frac{r}{M},$$

must be the same for all substances, with the same values of t and r; or, since r is not involved except as a factor, we must have

$$\frac{\frac{dp}{dt}}{M} = \mu, \qquad (4)$$

where μ depends only on t; from which we conclude the proposition which was to be proved.

22. The very remarkable theorem that $\frac{\frac{dp}{dt}}{M}$ must be the same for all substances at the same temperature was first given (although not in precisely the same terms) by Carnot, and demonstrated by him, according to the principles he adopted. We have now seen that its truth may be satisfactorily established without adopting the false part of his principles. Hence all Carnot's conclusions, and all conclusions derived by others from his theory, which depend merely on equation (3), require no modification when the dynamical theory is adopted. Thus, all the conclusions contained in Sections I., II., and III. of the Appendix to my "Account of Carnot's Theory," and in the paper immediately following it in the *Transactions*, entitled "Theoretical Considerations on the Effect of Pressure in Lowering the Freezing-point of Water," by my elder brother, still hold. Also, we see that Carnot's expression for the mechanical effect derivable from a given quantity of heat by means of a perfect

engine in which the range of temperatures is infinitely small, expresses truly the greatest effect which can possibly be obtained in the circumstances; although it is in reality only an infinitely small fraction of the whole mechanical equivalent of the heat supplied; the remainder being irrecoverably lost to man, and therefore "wasted," although not *annihilated*.

23. On the other hand, the expression for the mechanical effect obtainable from a given quantity of heat entering an engine from a "source" at a given temperature, when the range down to the temperature of the cold part of the engine or the "refrigerator" is finite, will differ most materially from that of Carnot; since, a finite quantity of mechanical effect being now obtained from a finite quantity of heat entering the engine, a finite fraction of this quantity must be converted from heat into mechanical effect. The investigation of this expression, with numerical determinations founded on the numbers deduced from Regnault's observations on steam, which are shown in Tables I. and II. of my former paper, constitutes the second part of the paper at present communicated.

PART II

On the Motive Power of Heat through Finite Ranges of Temperature

24. It is required to determine the quantity of work which a perfect engine, supplied from a source at any temperature, S, and parting with its waste heat to a refrigerator at any lower temperature, T, will produce from a given quantity, H, of heat drawn from the source.

25. We may suppose the engine to consist of an infinite number of perfect engines, each working within an infinitely small range of temperature, and arranged in a series of which the source of the first is the given source, the refrigerator of the last the given refrigerator, and the refrigerator of each intermediate engine is the source of that which follows it in the series. Each of these engines will, in any time, emit just as much less heat to its refrigerator than is supplied to it from its source, as is the equivalent of the mechanical work which it produces. Hence if t and $t + dt$ denote respectively the temperatures of the refrigerator and source of one of the inter-

mediate engines, and if q denote the quantity of heat which this engine discharges into its refrigerator in any time, and $q + dq$ the quantity which it draws from its source in the same time, the quantity of work which it produces in that time will be Jdq according to Prop. I., and it will also be $q\mu dt$ according to the expression of Prop. II., investigated in § 21 ; and therefore we must have

$$Jdq = q\mu dt.$$

Hence, supposing that the quantity of heat supplied from the first source, in the time considered is H, we find by integration

$$\log \frac{H}{q} = \frac{1}{J}\int_t^S \mu dt.$$

But the value of q, when $t = T$, is the final remainder discharged into the refrigerator at the temperature T; and therefore, if this be denoted by R, we have

$$\log \frac{H}{R} = \frac{1}{J}\int_T^S \mu dt, \tag{5}$$

from which we deduce

$$R = H\epsilon - \frac{1}{J}\int_T^S \mu dt. \tag{6}$$

Now the whole amount of work produced will be the mechanical equivalent of the quantity of heat lost; and, therefore, if this be denoted by W, we have

$$W = J(H - R), \tag{7}$$

and consequently, by (6),

$$W = JH\left\{1 - \epsilon^{-\frac{1}{J}\int_T^S \mu dt}\right\}. \tag{8}$$

26. To compare this with the expression $H\int_T^S \mu dt$, for the duty indicated by Carnot's theory,[*] we may expand the exponential in the preceding equation, by the usual series. We thus find

$$W = \left(1 - \frac{\theta}{1.2} + \frac{\theta^2}{1.2.3} - \text{etc.}\right).H\int_T^S \mu dt$$

where

$$\theta = \frac{1}{J}\int_T^S \mu dt \tag{9}$$

This shows that the work really produced, which always falls short of the duty indicated by Carnot's theory, approaches

---

* "Account," etc., Equation 7, § 31.

more and more nearly to it as the range is diminished ; and ultimately, when the range is infinitely small, is the same as if Carnot's theory required no modification, which agrees with the conclusion stated above in § 22.

27. Again, equation (8) shows that the real duty of a given quantity of heat supplied from the source increases with every increase of the range ; but that instead of increasing indefinitely in proportion to $\int_{T}^{S} \mu dt$, as Carnot's theory makes it do, it never reaches the value JH, but approximates to this limit, as $\int_{T}^{S} \mu dt$ is increased without limit. Hence Carnot's remark* regarding the practical advantage that may be anticipated from the use of the air-engine, or from any method by which the range of temperatures may be increased, loses only a part of its importance, while a much more satisfactory view than his of the practical problem is afforded. Thus we see that, although the full equivalent of mechanical effect cannot be obtained even by means of a perfect engine, yet when the actual source of heat is at a high enough temperature above the surrounding objects, we may get more and more nearly the whole of the admitted heat converted into mechanical effect, by simply increasing the effective range of temperature in the engine.

28. The preceding investigation (§ 25) shows that the value of Carnot's function, μ, for all temperatures within the range of the engine, and the absolute value of Joule's equivalent, J, are enough of data to calculate the amount of mechanical effect of a perfect engine of any kind, whether a steam-engine, an air-engine, or even a thermo-electric engine ; since, according to the axiom stated in § 12, and the demonstration of Prop. II., no inanimate material agency could produce more mechanical effect from a given quantity of heat, with a given available range of temperatures, than an engine satisfying the criterion stated in the enunciation of the proposition.

29. The mechanical equivalent of a thermal unit Fahrenheit, or the quantity of heat necessary to raise the temperature of a pound of water from 32° to 33° Fahr., has been determined by Joule in foot-pounds at Manchester, and the value which he gives as his best determination is 772.69. Mr. Rankine takes,

---

* "Account," etc. Appendix, Section iv.

as the result of Joule's determination, 772, which he estimates must be within $\frac{1}{300}$ of its own amount, of the truth. If we take $772\frac{5}{8}$ as the number, we find, by multiplying it by $\frac{9}{5}$, 1390 as the equivalent of the thermal unit Centigrade, which is taken as the value of J in the numerical applications contained in the present paper.

30. With regard to the determination of the values of μ for different temperatures, it is to be remarked that equation (4) shows that this might be done by experiments upon any substance whatever of indestructible texture, and indicates exactly the experimental data required in each case. For instance, by first supposing the medium to be air; and again, by supposing it to consist partly of liquid water and partly of saturated vapor, we deduce, as is shown in Part III. of this paper, the two expressions (6), given in § 30 of my former paper ("Account of Carnot's Theory"), for the value of μ at any temperature. As yet no experiments have been made upon air which afford the required data for calculating the value of μ through any extensive range of temperature; but for temperatures between 50° and 60° Fahr., Joule's experiments[*] on the heat evolved by the expenditure of a given amount of work on the compression of air kept at a constant temperature, afford the most direct data for this object which have yet been obtained; since, if Q be the quantity of heat evolved by the compression of a fluid subject to "the gaseous laws" of expansion and compressibility, W the amount of mechanical work spent, and t the constant temperature of the fluid, we have by (11) of § 49 of my former paper,

$$\mu = \frac{W.E}{Q(1+Et)}, \qquad (10)$$

which is in reality a simple consequence of the other expression for μ in terms of data with reference to air. Remarks upon the determination of μ by such experiments, and by another class of experiments on air originated by Joule, are reserved for a separate communication, which I hope to be able to make to the Royal Society on another occasion.

31. The second of the expressions (6), in § 30 of my former paper, or the equivalent expression (32), given below in the

---

[*] "On the Changes of Temperature produced by the Rarefaction and Condensation of Air," *Phil. Mag.*, vol. xxvi., May, 1845.

present paper, shows that μ may be determined for any temperature from determinations for that temperature of—

(1) The rate of variation with the temperature, of the pressure of saturated steam.

(2) The latent heat of a given weight of saturated steam.

(3) The volume of a given weight of saturated steam.

(4) The volume of a given weight of water.

The last mentioned of these elements may, on account of the manner in which it enters the formula, be taken as constant, without producing any appreciable effect on the probable accuracy of the result.

32. Regnault's observations have supplied the first of the data with very great accuracy for all temperatures between $-32°$ Cent. and $230°$.

33. As regards the second of the data, it must be remarked that all experimenters, from Watt, who first made experiments on the subject, to Regnault, whose determinations are the most accurate and extensive that have yet been made, appear to have either explicitly or tacitly assumed the same principle as that of Carnot which is overturned by the dynamical theory of heat; inasmuch as they have defined the " total heat of steam " as the quantity of heat required to convert a unit of weight of water at 0° into steam in the particular state considered. Thus Regnault, setting out with this definition for " the total heat of saturated steam," gives experimental determinations of it for the entire range of temperatures from 0° to 230° ; and he deduces the "latent heat of saturated steam " at any temperature, from the "total heat," so determined, by subtracting from it the quantity of heat necessary to raise the liquid to that temperature. Now, according to the dynamical theory, the quantity of heat expressed by the preceding definition depends on the manner (which may be infinitely varied) in which the specified change of state is effected : differing in different cases by the thermal equivalents of the differences of the external mechanical effect produced in the expansion. For instance, the total quantity of heat required to evaporate a quantity of water at 0°, and then, keeping it always in the state of saturated vapor,* bring it to the temperature 100°, cannot be so much as

* See below (Part III., § 58), where the " negative " specific heat of saturated steam is investigated. If the mean value of this quantity between 0° and 100° were —1.5 (and it cannot differ much from this) there would

three-fourths of the quantity required, first, to raise the temperature of the liquid to 100°, and then evaporate it at that temperature; and yet either quantity is expressed by what is generally received as a *definition* of the "total heat" of the saturated vapor. To find what it is that is really determined as "total heat" of saturated steam in Regnault's researches, it is only necessary to remark, that the measurement actually made is of the quantity of heat emitted by a certain weight of water in passing through a calorimetrical apparatus, which it enters as saturated steam, and leaves in the liquid state, the result being reduced to what would have been found if the final temperature of the water had been exactly 0°. For there being no external mechanical effect produced (other than that of sound, which it is to be presumed is quite inappreciable), the only external effect is the emission of heat. This must, therefore, according to the fundamental proposition of the dynamical theory, be independent of the intermediate agencies. It follows that, however the steam may rush through the calorimeter, and at whatever reduced pressure it may actually be condensed,* the heat emitted externally must be exactly the

be 150 units of heat emitted by a pound of saturated vapor in having its temperature raised (by compression) from 0° to 100°. The latent heat of the vapor at 0° being 606.5, the final quantity of heat required to convert a pound of water at 0° into saturated steam at 100°, in the first of the ways mentioned in the text, would consequently be 456.5, which is only about ⅝ of the quantity 637 found as "the total heat" of the saturated vapor at 100°, by Regnault.

* If the steam have to rush through a long fine tube, or through a small aperture within the calorimetrical apparatus, its pressure will be diminished before it is condensed; and there will, therefore, in two parts of the calorimeter be saturated steam at different temperatures (as, for instance, would be the case if steam from a high-pressure boiler were distilled into the open air); yet, on account of the heat developed by the fluid friction, which would be precisely the equivalent of the mechanical effect of the expansion wasted in the rushing, the heat measured by the calorimeter would be precisely the same as if the condensation took place at a pressure not appreciably lower than that of the entering steam. The circumstances of such a case have been overlooked by Clausius (Poggendorff's *Annalen*, 1850, No. 4, p. 510), when he expresses with some doubt the opinion that the latent heat of saturated steam will be truly found from Regnault's "total heat," by deducting "the sensible heat;" and gives as a reason that, in the actual experiments, the condensation must have taken place "under the same pressure, or nearly under the same pressure," as the evaporation. The question is not, *Did the condensation take place at a lower pressure than*

same as if the condensation took place under the full pressure
of the entering saturated steam ; and we conclude that *the
total heat,* as actually determined from his experiments by Regnault, is the quantity of heat that would be required, first to
raise the liquid to the specified temperature, and then to evaporate it at that temperature ; and that the principle on which
he determines the latent heat is correct.   Hence, through the
range of his experiments—that is, from 0° to 230°—we may consider the second of the data required for the calculation of μ
as being supplied in a complete and satisfactory manner.

34. There remains only the third of the data, or the volume
of a given weight of saturated steam, for which accurate experiments through an extensive range are wanting ; and no experimental researches bearing on the subject having been made
since the time when my former paper was written, I see no
reason for supposing that the values of μ which I then gave are
not the most probable that can be obtained in the present state
of science ; and, on the understanding stated in § 33 of that
paper, that accurate experimental determinations of the densities of saturated steam at different temperatures may indicate
considerable errors in the densities which have been assumed
according to the "gaseous laws," and may consequently render
considerable alterations in my results necessary, I shall still continue to use Table I. of that paper, which shows the values of
μ for the temperatures $\frac{1}{2}$, $1\frac{1}{2}$, $2\frac{1}{2}$...$230\frac{1}{2}$, or, the mean values
of μ for each of the 230 successive Centigrade degrees of the
air-thermometer above the freezing-point, as the basis of numerical applications of the theory.   It may be added, that any
experimental researches sufficiently trustworthy in point of accuracy, yet to be made, either on air or any other substance,
which may lead to values of μ differing from those, must be
admitted as proving a discrepancy between the true densities
of saturated steam, and those which have been assumed.*

that of the entering steam? but, *Did* Regnault *make the steam work an engine
in passing through the calorimeter, or was there so much noise of steam rushing through it as to convert an appreciable portion of the total heat into external mechanical effect?* And a negative answer to this is a sufficient reason
for adopting *with certainty* the opinion that the principle of his determination of the latent heat is correct.

* I cannot see that any hypothesis, such as that adopted by Clausius
fundamentally in his investigations on this subject, and leading, as he shows,
to determinations of the densities of saturated steam at different temper-

35. Table II. of my former paper, which shows the values of $\int_0^t \mu dt$ for $t = 1$, $t = 2$, $t = 3$, ... $t = 231$, renders the calculation of the mechanical effect derivable from a given quantity of heat by means of a perfect engine, with any given range included between the limits 0 and 231, extremely easy; since the quantity to be divided by J^* in the index of the exponential in the expression (8) will be found by subtracting the number in that table corresponding to the value of T, from that corresponding to the value of S.

36. The following tables show some numerical results which have been obtained in this way, with a few (contained in the lower part of the second table) calculated from values of $\int_0^t \mu dt$ estimated for temperatures above 230°, roughly, according to the rate of variation of that function within the experimental limits.

37. *Explanation of the Tables.*

Column I. in each table shows the assumed ranges.

Column II. shows ranges deduced by means of Table II. of the former paper, so that the value of $\int_T^S \mu dt$ for each may be the same as for the corresponding range shown in column I.

Column III. shows what would be the duty of a unit of heat if Carnot's theory required no modification (or the actual duty of a unit of heat with additions through the range, to compensate for the quantities converted into mechanical effect).

atures, which indicate enormous deviations from the gaseous laws of variation with temperature and pressure, is more probable, or is probably nearer the truth, than that the density of saturated steam does follow these laws as it is usually assumed to do. In the present state of science it would perhaps be wrong to say that either hypothesis is more probable than the other [or that the rigorous truth of either hypothesis is probable at all].

* It ought to be remarked, that as the unit of force implied in the determinations of μ is the weight of a pound of matter at Paris, and the unit of force in terms of which J is expressed is the weight of a pound at Manchester, these numbers ought in strictness to be modified so as to express the values in terms of a common unit of force; but as the force of gravity at Paris differs by less than $\frac{1}{1000}$ of its own value from the force of gravity at Manchester, this correction will be much less than the probable errors from other sources, and may therefore be neglected.

Column IV. shows the true duty of a unit of heat, and a comparison of the numbers in it with the corresponding numbers in Column III. shows how much the true duty falls short of Carnot's theoretical duty in each case.

Column VI. is calculated by the formula

$$R = \epsilon^{-\frac{1}{1390}\int_T^S \mu dt},$$

where $\epsilon = 2.71828$, and for $\int_T^S \mu dt$ the successive values shown in Column III. are used.

Column IV. is calculated by the formula

$$W = 1390\,(1 - R)$$

from the values of $1 - R$ shown in Column V.

38. *Table of the Motive Power of Heat.*

Range of Temperatures				III Duty of a unit of heat through the whole range	IV Duty of a unit of heat supplied from the source	V Quantity of heat converted into mechanical effect	VI Quantity of heat wasted
I		II					
S	T	S	T	$\int_T^S \mu dt$	W	$1 - R$	R
°	°	°	°	ft.-lbs.	ft.-lbs.		
1	0	31.08	30	4.960	4.948	.00356	.99644
10	0	40.86	30	48.987	48.1	.0346	.9654
20	0	51.7	30	96 656	93.4	.067	.933
30	0	62.6	30	143.06	136	.098	.902
40	0	73.6	30	188.22	176	.127	.873
50	0	84.5	30	232.18	214	.154	.846
60	0	95 4	30	274 97	249	.179	.821
70	0	106.3	30	316.64	283	.204	.796
80	0	117.2	30	357.27	315	.227	.773
90	0	128.0	30	396.93	345	.248	.752
100	0	138.8	30	435.69	374	.269	.731
110	0	149.1	30	473.62	401	.289	.711
120	0	160.3	30	510.77	427	.308	.692
130	0	171.0	30	547.21	452	.325	.675
140	0	181.7	30	582.98	476	.343	.657
150	0	192.3	30	618.14	499	.359	.641
160	0	203 0	30	652.74	521	.375	.625
170	0	213.6	30	686.80	542	.390	.610
180	0	224.2	30	720.39	562	.404	.596
190	0	190	0	753.50	582	.418	.582
200	0	200	0	786.17	600	.432	.568
210	0	210	0	818.45	619	.445	.555
220	0	220	0	850.34	636	.457	.542
230	0	230	0	881 87	653	.470	.530

39. *Supplementary Table of the Motive Power of Heat.*

Range of Temperatures				III Duty of a unit of heat through the whole range	IV Duty of a unit of heat supplied from the source	V Quantity of heat converted into mechanical effect	VI Quantity of heat wasted
1		II					
S	T	S	T	$\int_T^S \mu dt$	W	$1 - R$	R
°	°	°	°	ft.-lbs.	ft.-lbs.		
101.1	0	140	30	439.9	377	.271	.729
105.8	0	230	100	446.2	382	.275	.725
300	0	300	0	1099	757	.545	.455
400	0	400	0	1395	879	.632	.368
500	0	500	0	1690	979	.704	.296
600	0	600	0	1980	1059	.762	.238
∞	0	∞	0	∞	1390	1.000	.000

40. Taking the range 30° to 140° as an example suitable to the circumstances of some of the best steam-engines that have yet been made (see Appendix to " Account of Carnot's Theory," sec. v.), we find in Column III., of the supplementary table, 377 ft.-lbs. as the corresponding duty of a unit of heat instead of 440, shown in Column III., which is Carnot's theoretical duty. We conclude that the recorded performance of the Fowey-Consols engine in 1845, instead of being only 57½ per cent. amounted really to 67 per cent., or ⅔ of the duty of a perfect engine with the same range of temperature; and this duty being .271 (rather more than ¼) of the whole equivalent of the heat used ; we conclude further, that $\frac{1}{5.49}$, or 18 per cent. of the whole heat supplied was actually converted into mechanical effect by that steam-engine.

41. The numbers in the lower part of the supplementary table show the great advantage that may be anticipated from the perfecting of the air-engine, or any other kind of thermo-dynamic engine in which the range of the temperature can be increased much beyond the limits actually attainable in steam-engines. Thus an air-engine, with its hot part at 600°, and its cold part at 0° Cent., working with perfect economy, would convert 76 per cent. of the whole heat used into mechanical effect ; or working with such economy as has been estimated for the Fowey-Consols engine—that is, producing 67 per cent. of the theoretical duty corresponding to its range of temperature—

would convert 51 per cent. of all the heat used into mechanical effect.

42. It was suggested to me by Mr. Joule, in a letter dated December 9, 1848, that the true value of μ might be "inversely as the temperatures from zero;"* and values for various temperatures calculated by means of the formula,

$$\mu = J\frac{E}{1 + Et},\qquad(11)$$

were given for comparison with those which I had calculated from data regarding steam. This formula is also adopted by Clausius, who uses it fundamentally in his mathematical investigations. If μ were correctly expressed by it, we should have

$$\int_T^S \mu dt = J\,\log\frac{1 + ES}{1 + ET};$$

and therefore equations (1) and (2) would become

$$W = J\frac{S - T}{\dfrac{1}{E} + S},\qquad(12)$$

$$R = \frac{\dfrac{1}{E} + T}{\dfrac{1}{E} + S}.\qquad(13)$$

43. The reasons upon which Mr. Joule's opinion is founded, that the preceding equation (11) may be the correct expression

---

* If we take $\mu = k\dfrac{E}{1 + Et}$, where k may be any constant, we find

$$W = J\left(\frac{S - T}{\dfrac{1}{E} + S}\right)\frac{k}{J},$$

which is the formula I gave when this paper was communicated. I have since remarked that Mr. Joule's hypothesis implies essentially that the coefficient k must be as it is taken in the text, the mechanical equivalent of a thermal unit. Mr. Rankine, in a letter dated March 27, 1851, informs me that he has deduced, from the principles laid down in his paper communicated last year to this Society, an approximate formula for the ratio of the maximum quantity of heat converted into mechanical effect to the whole quantity expended, in an expansive engine of any substance, which, on comparison, I find agrees exactly with the expression (12) given in the text as a consequence of the hypothesis suggested by Mr. Joule regarding the value of μ at any temperature.—[April 4, 1851.]

for Carnot's function, although the values calculated by means of it differ considerably from those shown in Table I. of my former paper, form the subject of a communication which I hope to have an opportunity of laying before the Royal Society previously to the close of the present session.

PART III.

Applications of the Dynamical Theory to establish Relations between the Physical Properties of all Substances.

44. The two fundamental equations of the dynamical theory of heat, investigated above, express relations between quantities of heat required to produce changes of volume and temperature in any material medium whatever, subjected to a uniform pressure in all directions, which lead to various remarkable conclusions. Such of these as are independent of Joule's principle (expressed by equation (2) of § 20), being also independent of the truth or falseness of Carnot's contrary assumption regarding the permanence of heat, are common to his theory and to the dynamical theory; and some of the most important of them* have been given by Carnot himself, and other writers who adopted his principles and mode of reasoning without modification. Other remarkable conclusions on the same subject might have been drawn from the equation $\dfrac{dM}{dt} - \dfrac{dN}{dv} = 0$, expressing Carnot's assumption (of the truth of which experimental tests might have been thus suggested); but I am not aware that any conclusion deducible from it, not included in Carnot's expression for the motive power of heat through finite ranges of temperature, has yet been actually obtained and published.

45. The recent writings of Rankine and Clausius contain some of the consequences of the fundamental principle of the dynamical theory (expressed in the first fundamental proposition above) regarding physical properties of various substances; among which may be mentioned especially a very remarkable discovery regarding the specific heat of saturated steam (investigated also in this paper in § 58 below), made independently

* See above, § 22.

by the two authors, and a property of water at its freezing-point, deduced from the corresponding investigation regarding ice and water under pressure by Clausius; according to which he finds that, for each $\frac{1}{10}°$ Cent. that the solidifying point of water is lowered by pressure, its latent heat, which under atmospheric pressure is 79, is diminished by .081. The investigations of both these writers involve fundamentally various hypotheses which may be or may not be found by experiment to be approximately true; and which render it difficult to gather from their writings what part of their conclusions, especially with reference to air and gases, depend merely on the necessary principles of the dynamical theory.

46. In the remainder of this paper, the two fundamental propositions, expressed by the equations

$$\frac{dM}{dt} - \frac{dN}{dv} = \frac{1}{J}\frac{dp}{dt} \qquad \text{(2) of § 20}$$

and

$$M = \frac{1}{\mu}\cdot\frac{dp}{dt}, \qquad \text{(3) of § 21}$$

are applied to establish properties of the specific heats of any substance whatever; and then special conclusions are deduced for the case of a fluid following strictly the " gaseous laws " of density, and for the case of a medium consisting of parts in different states at the same temperature, as water and saturated steam, or ice and water.

47. In the first place it may be remarked, that by the definition of M and N in § 20, N must be what is commonly called the " specific heat at constant volume " of the substance, provided the quantity of the medium be the standard quantity adopted for specific heats, which, in all that follows, I shall take as the unit of weight. Hence the fundamental equation of the dynamical theory, (2) of § 20, expresses a relation between this specific heat and the quantities for the particular substance denoted by M and p. If we eliminate M from this equation, by means of equation (3) of § 21, derived from the expression of the second fundamental principle of the theory of the motive power of heat, we find

$$\frac{dN}{dv} = \frac{d\left(\frac{1}{\mu}\frac{dp}{dt}\right)}{dt} - \frac{1}{J}\frac{dp}{dt}, \qquad (14)$$

which expresses a relation between the variation in the specific heat at constant volume, of any substance, produced by an alteration of its volume at a constant temperature, and the variation of its pressure with its temperature when the volume is constant; involving a function, μ, of the temperature, which is the same for all substances.

48. Again, let K denote the specific heat of the substance under constant pressure. Then, if dv and dt be so related that the pressure of the medium, when its volume and temperature are $v + dv$ and $t + dt$ respectively, is the same as when they are v and t—that is, if

$$0 = \frac{dp}{dv}dv + \frac{dp}{dt}dt,$$

we have

$$Kdt = Mdv + Ndt.$$

Hence we find

$$M = \frac{-\dfrac{dp}{dv}}{\dfrac{dp}{dt}}(K - N), \qquad (15)$$

which merely shows the meaning in terms of the two specific heats, of what I have denoted by M. Using in this for M its value given by (3) of § 21, we find

$$K - N = \frac{\left(\dfrac{dp}{dt}\right)^2}{\mu \times -\dfrac{dp}{dv}}, \qquad (16)$$

an expression for the difference between the two specific heats, derived without hypothesis from the second fundamental principle of the theory of the motive power of heat.

49. These results may be put into forms more convenient for use, in applications to liquid and solid media, by introducing the notation :

$$\left.\begin{aligned} \kappa &= v \times -\frac{dp}{dv} \\[2mm] e &= \frac{1}{\kappa}\frac{dp}{dt} \end{aligned}\right\}, \qquad (17)$$

where κ will be the reciprocal of the compressibility, and e the coefficient of expansion with heat.

Equations (14), (16), and (3) thus become

$$\frac{dN}{dv} = \frac{d\left(\frac{\kappa e}{\mu}\right)}{dt} - \frac{\kappa e}{J}, \tag{18}$$

$$K - N = v\frac{\kappa e^2}{\mu}, \tag{19}$$

$$M = \frac{1}{\mu} \cdot \kappa e; \tag{20}$$

the third of these equations being annexed to show explicitly the quantity of heat developed by the compression of the substance kept at a constant temperature. Lastly, if θ denote the rise in temperature produced by a compression from $v + dv$ to v before any heat is emitted, we have

$$\theta = \frac{1}{N} \cdot \frac{\kappa e}{\mu} \cdot dv = -\frac{\kappa e}{\mu K - v\kappa e^2} dv. \tag{21}$$

50. The first of these expressions for θ shows that, when the substance contracts as its temperature rises (as is the case, for instance, with water between its freezing-point and its point of maximum density), its temperature would become lowered by a sudden compression. The second, which shows in terms of its compressibility and expansibility exactly how much the temperature of any substance is altered by an infinitely small alteration of its volume, leads to the approximate expression

$$\theta = \frac{\kappa e}{\mu K},$$

if, as is probably the case, for all known solids and liquids, e be so small that $e \cdot v\kappa e$ is very small compared with μK.

51. If, now, we suppose the substance to be a gas, and introduce the hypothesis that its density is strictly subject to the "gaseous laws," we should have, by Boyle and Mariotte's law of compression,

$$\frac{dp}{dv} = -\frac{p}{v}, \tag{22}$$

and by Dalton and Gay-Lussac's law of expansion,

$$\frac{dv}{dt} = \frac{Ev}{1 + Et}; \tag{23}$$

from which we deduce

$$\frac{dp}{dt} = \frac{Ep}{1 + Et}.$$

Equation (14) will consequently become

$$\frac{dN}{dv} = \frac{d\left\{\frac{Ep}{\mu(1+Et)} - \frac{p}{J}\right\}}{dt}, \tag{24}$$

a result peculiar to the dynamical theory and equation (16),

$$K - N = \frac{E'pv}{\mu(1-Et)^{2}}, \tag{25}$$

which agrees with the result of § 53 of my former paper.

If V be taken to denote the volume of the gas at the temperature 0° under unity of pressure, (25) becomes

$$K - N = \frac{E^{2}V}{\mu(1+Et)}. \tag{26}$$

52. All the conclusions obtained by Clausius, with reference to air or gases, are obtained immediately from these equations by taking

$$\mu = J\frac{E}{1+Et},$$

which will make $\frac{dN}{dv} = 0$, and by assuming, as he does, that N, thus found to be independent of the density of the gas, is also independent of its temperature.

53. As a last application of the two fundamental equations of the theory, let the medium with reference to which M and N are defined consist of a weight $1 - x$ of a certain substance in one state, and a weight x in another state at the same temperature, containing more latent heat. To avoid circumlocution and to fix the ideas, in what follows we may suppose the former state to be liquid and the latter gaseous; but the investigation, as will be seen, is equally applicable to the case of a solid in contact with the same substance in the liquid or gaseous form.

54. The volume and temperature of the whole medium being, as before, denoted respectively by v and t, we shall have

$$\lambda(1-x) + \gamma x = v, \tag{27}$$

if λ and γ be the volumes of unity of weight of the substance in the liquid and the gaseous states respectively: and p, the pressure, may be considered as a function of t, depending solely on the nature of the substance. To express M and N for this mixed medium, let L denote the latent heat of a unit of weight of the vapor, c the specific heat of the liquid, and h the specific heat of the vapor when kept in a state of saturation. We shall have

$$Mdv = L\frac{dx}{dv}dv,$$

$$Ndt = c(1-x)dt + hxdt + L\frac{dx}{dt}dt.$$

Now, by (27), we have

$$(\gamma - \lambda)\frac{dx}{dv} = 1, \tag{28}$$

and

$$(\gamma - \lambda)\frac{dx}{dt} + (1-x)\frac{d\lambda}{dt} + x\frac{d\gamma}{dt} = 0. \tag{29}$$

Hence

$$M = \frac{L}{\gamma - \lambda}, \tag{30}$$

$$N = c(1-x) + hx - L\frac{(1-x)\frac{d\lambda}{dt} + x\frac{d\gamma}{dt}}{\gamma - \lambda}. \tag{31}$$

55. The expression of the second fundamental proposition in this case becomes, consequently,

$$\mu = \frac{(\gamma - \lambda)\frac{dp}{dt}}{L}, \tag{32}$$

which agrees with Carnot's original result, and is the formula that has been used (referred to above in § 31) for determining μ by means of Regnault's observations on steam.

56. To express the conclusion derivable from the first fundamental proposition, we have, by differentiating the preceding expressions for M and N with reference to t and v respectively,

$$\frac{dM}{dv} = \frac{1}{\gamma - \lambda} \cdot \frac{dL}{dt} - \frac{L}{(\gamma - \lambda)^2} \cdot \frac{d(\gamma - \lambda)}{dt},$$

$$\frac{dN}{dt} = \left(h - c - L\frac{\frac{d\gamma}{dt} - \frac{d\lambda}{dt}}{\gamma - \lambda}\right)\frac{dx}{dv}$$

$$= \left\{\frac{h-c}{\gamma - \lambda} - \frac{L}{(\gamma - \lambda)^2}\right\}\frac{d(\gamma - \lambda)}{dt}.$$

Hence equation (2) of § 20 becomes

$$\frac{\frac{dL}{dt} + c - h}{\gamma - \lambda} = \frac{1}{J}\frac{dp}{dt}. \tag{33}$$

Combining this with the conclusion (32) derived from the second fundamental proposition, we obtain ·

$$\frac{dL}{dt} + c - h = \frac{L\mu}{J}. \tag{34}$$

The former of these equations agrees precisely with one which was first given by Clausius, and the preceding investigation is substantially the same as the investigation by which he arrived at it. The second differs from another given by Clausius only in not implying any hypothesis as to the form of Carnot's function μ.

57. If we suppose μ and L to be known for any temperature, equation (32) enables us to determine the value of $\frac{dp}{dt}$ for that temperature ; and thence deducing a value of dt, we have

$$dt = \frac{\gamma - \lambda}{\mu L} dp, \tag{35}$$

which shows the effect of pressure in altering the "boiling-point" if the mixed medium be a liquid and its vapor, or the melting-point if it be a solid in contact with the same substance in the liquid state. This agrees with the conclusion arrived at by my elder brother in his "Theoretical Investigation of the Effect of Pressure in Lowering the Freezing-point of Water." His result, obtained by taking as the value for μ that derived from Table I. of my former paper for the temperature $0°$, is that the freezing-point is lowered by $.0075°$ Cent. by an additional atmosphere of pressure. Clausius, with the other data the same, obtains $.00733°$ as the lowering of temperature by the same additional pressure, which differs from my brother's result only from having been calculated from a formula which implies the hypothetical expression $J\dfrac{E}{1 + Et}$ for μ. It was by applying equation (33) to determine $\frac{dL}{dt}$ for the same case that Clausius arrived at the curious result regarding the latent heat of water under pressure mentioned above (§ 45).

58. Lastly, it may be remarked that every quantity which appears in equation (33), except h, is known with tolerable accuracy for saturated steam through a wide range of temperature ; and we may therefore use this equation to find h, which has never yet been made an object of experimental research.

Thus we have

$$- h = \frac{\gamma - \lambda}{J} \frac{dp}{dt} - \left(\frac{dL}{dt} + c \right).$$

For the value of γ the best data regarding the density of saturated steam that can be had must be taken. If for different temperatures we use the same values for the density of saturated steam (calculated according to the gaseous laws, and Regnault's observed pressure from $\frac{1}{1693.5}$, taken as the density at 100°), the values obtained for the first term of the second member of the preceding equation are the same as if we take the form

$$- h = \frac{L\mu}{J} - \left(\frac{dL}{dt} + c \right)$$

derived from (34), and use the values of μ shown in Table I. of my former paper. The values of $-h$ in the second column in the following table have been so calculated, with, besides, the following data afforded by Regnault from his observations on the total heat of steam, and the specific heat of water

$$\frac{dL}{dt} + c = .305,$$

$$L = 606.5 + .305l - (.00002l^2 + .000000l^3).$$

The values of $-h$ shown in the third column are those derived by Clausius from an equation which is the same as what (34) would become if $J \dfrac{E}{1 + El}$ were substituted for μ.

t.	$-h$ according to Table I. of "Account of Carnot's Theory"	$-h$ according to Clausius
0	1.863	1.916
50	1.479	1.465
100	1.174	1.133
150	0.951	0.879
200	0.780	0.676

59. From these results it appears, that through the whole range of temperatures at which observations have been made, the value of h is negative; and, therefore, if a quantity of saturated vapor be compressed in a vessel containing no liquid water, heat must be continuously abstracted from it in order that it may remain saturated as its temperature rises; and conversely, if a quantity of saturated vapor be allowed to expand

in a closed vessel, heat must be supplied to it to prevent any
part of it from becoming condensed into the liquid form as the
temperature of the whole sinks. This very remarkable conclusion was first announced by Mr. Rankine, in his paper communicated to this Society on the 4th of February last year. It
was discovered independently by Clausius, and published in
his paper in Poggendorff's *Annalen* in the months of April and
May of the same year.

60. It might appear at first sight, that the well-known fact
that steam rushing from a high-pressure boiler through a small
orifice into the open air does not scald a hand exposed to it, is
inconsistent with the proposition, that steam expanding from
a state of saturation must have heat given to it to prevent any
part from becoming condensed; since the steam would scald
the hand unless it were dry, and consequently above the boiling-point in temperature. The explanation of this apparent
difficulty, given in a letter which I wrote to Mr. Joule last October, and which has since been published in the *Philosophical
Magazine*, is, that the steam in rushing through the orifice produces mechanical effect which is immediately wasted in fluid
friction, and consequently reconverted into heat; so that the
issuing steam at the atmospheric pressure would have to part
with as much heat to convert it into water at the temperature
100° as it would have had to part with to have been condensed
at the high pressure and then cooled down to 100°, which for
a pound of steam initially saturated at the temperature t is, by
Regnault's modification of Watt's law, .305 $(t - 100°)$ more heat
than a pound of saturated steam at 100° would have to part
with to be reduced to the same state; and the issuing steam
must therefore be above 100° in temperature, and dry.

Part IV

*On a Method of discovering experimentally the Relation between
the Mechanical Work spent and the Heat produced by
the Compression of a Gaseous Fluid.**

61. The important researches of Joule on the thermal circumstances connected with the expansion and compression of

*From the *Transactions of the Royal Society of Edinburgh*, vol. xx., part 2,
April 17, 1851.

air, and the admirable reasoning upon them expressed in his paper* "On the Changes of Temperature produced by the Rarefaction and Condensation of Air," especially the way in which he takes into account any mechanical effect that may be externally produced, or internally lost, in fluid friction, have introduced an entirely new method of treating questions regarding the physical properties of fluids. The object of the present paper is to show how, by the use of this new method, in connection with the principles explained in my preceding paper, a complete theoretical view may be obtained of the phenomena experimented on by Joule ; and to point out some of the objects to be attained by a continuation and extension of his experimental researches.

62. The Appendix to my "Account of Carnot's Theory"† contains a theoretical investigation of the heat developed by the compression of any fluid fulfilling the laws‡ of Boyle and Mariotte and of Dalton and Gay-Lussac. It has since been shown that that investigation requires no modification when the dynamical theory is adopted, and therefore the formula obtained as the result may be regarded as being established for a fluid of the kind assumed, independently of any hypothesis whatever. We may obtain a corresponding formula applicable to a fluid not fulfilling the gaseous laws of density, or to a solid pressed uniformly on all sides, in the following manner:

63. Let Mdv be the quantity of heat absorbed by a body kept at a constant temperature t, when its volume is increased from v to $v + dv$; let p be the uniform pressure which it experiences from without, when its volume is v and its temperature t; and let $p + \dfrac{dp}{dt} dt$ denote the value p would acquire if the temperature were raised to $t + dt$, the volume remaining unchanged. Then, by equation (3) of § 21 of my former paper, derived from Clausius's extension of Carnot's theory, we have

$$M = \frac{1}{\mu} \cdot \frac{dp}{dt}, \qquad (a)\S$$

---

* *Philosophical Magazine*, May, 1845, vol. xxvi., p. 369.

† *Transactions*, vol. xvi., part 5.

‡ To avoid circumlocution, these laws will, in what follows, be called simply the *gaseous laws*, or the *gaseous laws* of density.

§ Throughout this paper, formulæ which involve no hypothesis whatever are marked with italic letters ; formulæ which involve Boyle's and Dalton's laws are marked with Arabic numerals ; and formulæ involving, besides, Mayer's hypothesis, are marked with Roman numerals.

where μ denotes Carnot's *function*, the same for all substances at the same temperature.

Now let the substance expand from any volume V to V', and, being kept constantly at the temperature t, let it absorb a quantity, H, of heat. Then

$$H = \int_V^{V'} M dv = \frac{1}{\mu}\frac{d}{dt}\int_V^{V'} p dv. \qquad (b)$$

But if W denote the mechanical work which the substance does in expanding, we have

$$W = \int_V^{V'} p dv, \qquad (c)$$

and therefore

$$H = \frac{1}{\mu}\frac{dW}{dt}. \qquad (d)$$

This formula, established without any assumption admitting of doubt, expresses the relation between the heat developed by the compression of any substance whatever, and the mechanical work which is required to effect the compression, as far as it can be determined without hypothesis by purely theoretical considerations.

64. The preceding formula leads to that which I formerly gave for the case of fluids subject to the gaseous laws; since for such we have

$$pv = p_0 v_0 (1 + Et), \qquad (1)$$

from which we deduce, by (c),

$$W = p_0 v_0 (1 + Et) \log\frac{V'}{V}, \qquad (2)$$

and

$$\frac{dW}{dt} = E p_0 v_0 . \log\frac{V'}{V} = \frac{E}{1 + Et} W ; \qquad (3)$$

and therefore, by (d),

$$H = \frac{E}{\mu(1 + Et)} W, \qquad (4)$$

which agrees with equation (11) of § 49 of the former paper.

65. Hence we conclude, that the heat evolved by any fluid fulfilling the gaseous laws is proportional to the work spent in compressing it at any given constant temperature ; but that the quantity of work required to produce a unit of heat is not constant for all temperatures, unless Carnot's function for different temperatures vary inversely as $1 + Et$; and that it is not the simple mechanical equivalent of the heat, as it was unwar-

rantably* assumed by Mayer to be, unless this function have precisely the expression

$$\mu = J \cdot \frac{E}{1 + Et}.$$ (I.)

This formula was suggested to me by Mr. Joule, in a letter dated December 9, 1848, as probably a true expression for μ, being required to reconcile the expression derived from Carnot's theory (which I had communicated to him) for the heat evolved in terms of the work spent in the compression of a gas, with the hypothesis that the latter of these is exactly the mechanical equivalent of the former, which he had adopted in consequence of its being, at least approximately, verified by his own experiments. This, which will be called Mayer's hypothesis, from its having been first assumed by Mayer, is also assumed by Clausius without any reason from experiment ; and an expression for μ the same as the preceding, is consequently adopted by him as the foundation of his mathematical deductions from elementary reasoning regarding the motive power of heat. The preceding formulæ show, that if it be true at a particular temperature for any one fluid fulfilling the gaseous laws, it must be true for every such fluid at the same temperature.

[*The remaining sections are omitted. They deal with the experimental verification of Mayer's hypothesis, with the mechanical energy of a fluid, and with the applications of thermodynamics to electrical phenomena.*]

BIOGRAPHICAL SKETCH

WILLIAM THOMSON, now Lord Kelvin, was born at Belfast in June, 1824. His father, James Thomson, an eminent mathematician and student of science, removed in 1832 to Glasgow, where he occupied a position as professor in the university. His son studied under him at Glasgow and at St. Peter's College, Cambridge, and was graduated in 1845 at Cambridge, as Second Wrangler and Smith's Prizeman. He was called to the

---

* In violation of Carnot's important principle, that thermal agency and mechanical effect, or mechanical agency and thermal effect, cannot be regarded in the simple relation of cause and effect, when any other effect, such as the alteration of the density of a body, is finally concerned.

Chair of Natural Philosophy at Glasgow University in 1846, and he has since been connected with that university. On the completion of the Atlantic Cable in 1866, to the success of which Thomson had materially contributed by his study of the theoretical questions involved and by his numerous inventions, he was knighted, and on New-Year's day, 1892, he was raised to the peerage as Lord Kelvin.

His contributions to physics range over the whole domain of the science. The most important of these relate to electricity and magnetism, and to the mechanical theory of heat. Of great value also are the improvements made by him in electrical measurements, by his invention of the mirror galvanometer and of the various electrometers which bear his name. His work is characterized by its great versatility and by a peculiarly happy combination of profound theoretical knowledge and power of analysis with the ability to invent and execute important experimental investigations.

BIBLIOGRAPHY

Books of Reference

Clausius.	*Die Mechanische Wärmetheorie.*
Maxwell.	*Theory of Heat.*
Tait.	*Thermodynamics.*
Verdet.	*Théorie Mécanique de la Chaleur.*
Rankine.	*The Steam-engine.*
Briot.	*La Chaleur.*
Poincaré.	*Thermodynamique.*
Planck.	*Thermodyamik.*
Mach.	*Die Principien der Wärmelehre.*
Tait.	*Recent Advances in Physical Science.*

Articles

On the Validity of the Second Law

Rankine.	*Phil. Mag.* (IV.), 4, p. 358.
Holtzmann.	Pogg. *Ann.*, 82, p. 445.
Dechcr.	Dingler's *Pol. Jour.*, 148, pp. 1, 81, 161, 241.
Hirn.	*Cosmos*, 22, pp. 283, 413. (Also in *Exposition Analytique et Expérimentale de la Théorie Mécanique de la Chaleur.*)
Wand.	*Carl's Repertorium*, 4, p. 281, 369.
Tait.	*Phil. Mag.* (IV.), 43.

The foregoing articles contain criticisms of Clausius's form of the Second Law. They are answered by Clausius in Appendices to his book, *Die Mechanische Wärmetheorie*, 2d ed., 1876.

On Mechanical Analogies to the Second Law

Clausius.	Pogg. *Ann.*, 93, p. 481.
	Phil. Mag. (IV.), 35, p. 405.
	Pogg. *Ann.*, 142, p. 433.
	Ibid., 146, p. 585.
Rankine.	*Phil. Mag.* (IV.), 30, p. 241.

Boltzmann. *Ber. der Wiener Akad.*, 53, II., p. 188, 195.
            Ibid., 68, II., pp. 526, 712.
            Ibid., 76, II., p. 373.
            Ibid., 78, II., pp. 1, 733.
Szily.   Pogg. *Ann.*, 145, pp. 295, 302 ; 149, p. 74.
Burbury.   *Phil. Mag.* (V.), 1, p. 61.

Of these the later papers by Boltzmann and that by Burbury discuss the relation between the Second Law and the Theory of Probability. The same subject is developed by Boltzmann in his book, *Vorlesungen über Gastheorie*, 2d part.